PEST CONTROL: An Assessment of Present and Alternative Technologies

VOLUME V

Pest Control and Public Health

The Report of the Public Health Study Team
Study on Problems of Pest Control
Environmental Studies Board
National Research Council

NATIONAL ACADEMY OF SCIENCES
Washington, D.C. 1976

This is Volume V of *Pest Control: An Assessment of Present and Alternative Technologies*. The other volumes are:

NOTICE: The project that is the subject of this report was approved by the Governing Board of the National Research Council, whose members are drawn from the Councils of the National Academy of Sciences, the National Academy of Engineering, and the Institute of Medicine. The members of the Committee responsible for the report were chosen for their special competences and with regard for appropriate balance.

This report has been reviewed by a group other than the authors according to procedures approved by a Report Review Committee consisting of members of the National Academy of Sciences, the National Academy of Engineering, and the Institute of Medicine.

Library of Congress Catalog Card No. 75-45777
International Standard Book No. 0-309-02414-5 (this volume)
0-309-02409-9 (entire set)

Available from

Printing and Publishing Office
National Academy of Sciences
2101 Constitution Avenue, N.W.
Washington, D.C. 20418

Printed in the United States of America

PREFACE

This study originated in discussions in the Executive Offices of the National Academy of Sciences in March 1971. The Environmental Studies Board (ESB)* subsequently continued the discussions and appointed a Planning Committee to develop a detailed proposal. The Planning Committee, sponsored by funds from the Scaife Family Charitable Trusts, submitted its report to the ESB in April 1972. The ESB approved the report and appointed Dr. Donald Kennedy of Stanford University as Chairman of the Executive Committee of the Study on Problems of Pest Control: A Technology Assessment. Financial support for the study was secured from the U.S. Department of Agriculture, the U.S. Environmental Protection Agency, and the Ford Foundation. We thank these public and private agencies for their assistance.

The Executive Committee was charged with assessing the diverse effects on society that arise from efforts to control pests and to evaluate long-range alternatives available for accomplishing pest control. The charge also called explicitly for an examination of social and institutional factors important to the conduct of pest control

*The National Academy of Sciences complex (National Academy of Sciences, National Academy of Engineering, Institute of Medicine) functions as official but independent advisor to the federal government in matters relating to science and technology through its operating agency, the National Research Council. The National Research Council comprises four Assemblies and four Commissions, each of which deals with specific disciplines or societal problems. The Commission on Natural Resources, formed in 1973, oversees the work of five boards, among them the Environmental Studies Board. Thus the pest control study, having originated as described above, completed its activities under the aegis of the Commission on Natural Resources.

practices, factors that must be considered in any recommendations for changes in pest control policy. The Planning Committee and the Executive Committee concluded that the most effective way to examine pest control problems would be through a series of case studies. Four such studies were selected. Three of them centered on specific agroecosystems: cotton, corn and soybeans, and timber. The fourth case study dealt with issues related to the control of pests of public health significance in the United States and abroad. Each case study examined the pest species involved, current control measures, and the benefits and costs of controlling the pests by various techniques.

This volume presents the report of the Public Health Study Team. The specifics of the study charge and a description of the scope of the report are set forth in the Introduction. It should be noted here, however, that constraints of time and other resources prevented the study team from responding to the full range of its charge.

The case history approach was used in many areas because adequate data on which to base a critical analysis of the problem are often not available, and because almost all public health problems have unique characteristics that make generalizations difficult if not impossible. Because of its broad charge, the study team had to struggle with the interacting complexities of the need for pesticides to increase food supplies and protect public health and the potential negative impacts on human health through pesticide intoxication, the interactions in uses of pesticides for agriculture and for public health, and the complexities of improved nutrition and health as they relate to population growth. The study team does not consider that its investigation and conclusions have resolved these complex problems, but it hopes that this document will provide additional insight into them and will assist those who must deal with them directly.

The reports of the Executive Committee and the study teams are all available from the National Academy of Sciences in five volumes under the general title *Pest Control: An Assessment of Present and Alternative Technologies*. Volume I comprises the report of the Executive Committee and all the summaries and recommendations of the four study team reports. The report of the Executive Committee, in addition to drawing on the findings of the individual study teams, covers other areas not specifically dealt with by those teams. The remaining four volumes are *Volume II: Corn/Soybeans Pest Control; Volume III: Cotton Pest Control; Volume IV: Forest Pest Control; and Volume V: Pest Control and Public Health.*

ACKNOWLEDGMENTS

The committee met five times: once in New Orleans, Louisiana; once in Berkeley, California; and three times in Washington, D.C. Working papers on special topics were developed by the study team members or by our staff assistant, Ms. Judith Hough, as the basis for discussion in these meetings. At each meeting, experts in various aspects of the study team's charge who were asked to join the discussions added immeasurably to the scope and effectiveness of the team's deliberations.

A. Ralph Barr, Professor of Medical Entomology, School of Public Health, University of California, Los Angeles
Ernest C. Bay, Head, Department of Entomology, University of Maryland
George T. Carmichael, Administrative Director, Division of Mosquito Control, New Orleans
Harold C. Chapman, Location Leader, Gulf Coast Mosquito Research Laboratory, Lake Charles, Louisiana
Louis A. Falcon, Associate Insect Pathologist, University of California, Berkeley
Arthur F. Geib,* Manager, Kern Mosquito Abatement District, Bakersfield, California
George P. Georghiou, Professor of Entomology, University of California, Riverside
Eugene E. Kauffman, Sutter-Yuba Mosquito Abatement District, Yuba City
Duane S. Mikkelsen, Professor of Agronomy, University of California, Davis
Thomas H. Milby, California State Department of Public Health, Sacramento

*Deceased.

Carl J. Mitchell, Extension Specialist in Mosquito Research, Agricultural Extension Service, University of California, Davis

Thomas D. Mulhern, California State Department of Public Health, Bureau of Vector Control and Solid Waste Management, Sacramento

Mir S. Mulla, Professor of Entomology, University of California, Riverside

W. Donald Murray, Secretary-Treasurer, American Mosquito Control Association

John Osmun, Director, Operations Division, Environmental Protection Agency Pesticide Programs, Washington, D.C.

Richard F. Peters, Chief, Bureau of Vector Control and Solid Waste Management, California State Department of Public Health, Sacramento

William C. Reeves, Professor of Epidemiology, School of Public Health, University of California, Berkeley

Edgar Smith, Office of Health, U.S. Agency for International Development, Washington, D.C.

Robert C. Spear, Assistant Professor of Environmental Health Science, University of California, Berkeley

C. Dayton Steelman, Department of Entomology, Baton Rouge, Louisiana

Robert K. Washino, Associate Entomologist in Experiment Station, University of California, Davis

Donald E. Weidhaas, Director, Southern Region, Florida-Antilles Area, Insects Affecting Man Research Laboratory, Gainesville, Florida

In addition, one of the members of the committee, Dr. Maurice Provost, consulted the directors of 13 mosquito abatement districts for detailed information on their programs. These men responded generously to his request for information, and the facts they supplied form the basis for much of the section on organized mosquito control in the United States (see Chapter 2). Those who provided information were

C. G. Alvarez, Desplaines Valley Mosquito Abatement District, Illinois

A. W. Buzicky,* Metropolitan Mosquito Control District, Minnesota

G. T. Carmichael, New Orleans Division of Mosquito Control, Louisiana

*Deceased.

G. C. Collett, Salt Lake City Mosquito Abatement District, Utah
T. C. Fultz, Chatham County Mosquito Control Commission, Georgia
W. E. Hazeltine, Butte County Mosquito Abatement District, California
J. H. Heidt, Dade County Mosquito Control Division, Florida
P. J. Hunt, East Volusia Mosquito Control District, Florida
J. H. Kimball, Orange County Mosquito Abatement District, California
E. B. Ladd, Clackamas County Vector Control District, Oregon
F. H. Lesser, Ocean County Mosquito Extermination Commission, New Jersey
W. D. Murray, Delta Vector Control District, California
J. D. Pauly, South Cook County Mosquito Abatement District, Illinois

Judith Hough, of the National Research Council staff, attended all meetings of the study team, assisted in arrangements, and participated in the preparation of the final report. Dr. John Perkins served as principal staff officer for the study as a whole, and Bette Lou Fields took over his responsibilities in its final phases. Linda Jones, Lucette Comer, Mary Dong, Susan Lesser, Christina Neumann and Estelle Miller provided able and tireless secretarial support.

The report that follows, much as it owes to the expert advice, cooperation, and support of those listed above, remains the sole responsibility of the study team.

PUBLIC HEALTH STUDY TEAM

Ray F. Smith, *Chairman*
Professor of Entomology
University of California, Berkeley
Berkeley, California

A. W. A. Brown
Director
Department of Entomology
Michigan State University
East Lansing, Michigan

H. Rouse Caffey
Associate Director
Louisiana Agricultural Experiment Station
Baton Rouge, Louisiana

John E. Davies
Professor and Chairman
Department of Epidemiology and Public Health
University of Miami School of Medicine
Miami, Florida

Walter P. Falcon
Director
Food Research Institute
Stanford University
Stanford, California

Maurice W. Provost
Director
Florida Medical Entomology Laboratory
Florida Division of Health
Vero Beach, Florida

Lloyd E. Rozeboom
Professor of Medical Entomology
The Johns Hopkins University
Baltimore, Maryland

John E. Scanlon
Professor of Medical Zoology
The University of Texas Health Science Center at Houston School of Public Health
Houston, Texas

Thomas H. Weller
Richard Pearson Strong Professor of Tropical Public Health and Chairman of the Department
Harvard School of Public Health
Boston, Massachusetts

CONTENTS

Appendixes

SUMMARY AND RECOMMENDATIONS

Vector-borne disease is now, as in the past, a major deterrent in some areas to human settlement and agricultural development. Nevertheless, the use of pesticides together with other tactics to control arthropods and other organisms of public health significance has had tremendous impact during the past quarter century on improving the health and well-being of man throughout the world. In the past decade, progress toward the elimination of disease has had setbacks in areas, in large part because of environmental and social change. The initial successes with malaria control utilizing DDT are not being repeated in all areas. In some of the problem areas, the habits of man himself in relation to those of the vector are responsible for continued transmission. Problems are also developing with other diseases of man (see Chapter 1).

An attack on the disease vector is not the only means of reducing or eliminating an arthropod-borne disease. The other approaches are protection by vaccines and prophylaxis by suppressive chemotherapy to minimize opportunities for infection of specific vectors. However, analysis of these indicates that the majority of the important human vector-borne diseases cannot be prevented by vaccines or by chemotherapy. Of necessity, their control is based on reduction of vector sources and reduction of contacts between vector and man (see Chapter 1).

Rapid changes in density and distribution of people, in economic and social values, and in some instances in the behavior of public health pests and the characteristics of disease require continued reevaluation and study of the strategies and tactics of pest control.

The control of arthropod vectors of disease or other pests of public health should be attempted only with recognition, and insofar as possible an intimate knowledge, of the significance of the ecology and behavior of the target species and the epidemiology of the disease and with appreciation for environmental values that may be depreciated.

As every vector control problem is unique, each will require evaluation by persons trained in epidemiology, entomology, engineering, ecology, and economics to determine whether control should be undertaken, and, if so, with which tactics and strategy. The benefits of pest control, whether by application of a pesticide, drainage to reduce a vector source, or some other action, must be weighed against possible unfavorable impacts on nontarget organisms and the quality of the environment (see Chapter 4).

The very complexity of these broad ecological approaches to public health pest control and the additional burden of economic impacts create new responsibilities for pest control practitioners. Consequently, their breadth and level of training must be increased and raised.

The study team therefore recommends that increased educational opportunities be made available in all aspects of integrated control of arthropods of public health significance.

Mosquitoes are by far the most important arthropods subject to control for general public health reasons in the United States. It is probable that 80-90 percent of the entire public health pest control effort is directed toward mosquitoes. The most important mosquito-borne diseases in the United States have largely disappeared, but their vectors remain. The United States is in what might be termed a receptive state for several important arthropod-borne diseases: of the elements required to produce an epidemic of several arthropod-borne diseases may be present in this country at any time (see Chapter 2).

The study team recommends that to be prepared for any epidemic of arthropod-borne disease, the highly developed reporting system of the health departments of cities, counties, and states, and extensive surveillance under the general supervision of the Center for Disease Control, United States Public Health Service, be maintained.

The majority of Americans living in coastal and other areas subject to an above-average mosquito nuisance problem enjoy the benefits of organized mosquito control. The sources of mosquito annoyance in different mosquito control districts reflect geography, terrain, and degree of urbanization. Unorganized mosquito control, whether by public or private agencies, forms a small portion of all mosquito control operations. It suffers, however, from lack of technical information and regulation.

The study team recommends that, because these operations are aimed primarily at adult arthropods, and since control is achieved by use of fogs and sprays containing toxicants that may affect human beings, some form of regulation beyond general licensing should be required.

Areawide public health pest control can be accomplished by either the public or the private sector of the economy. The fact that it requires manipulation of everyone's air, water, and land, however, bestows a predominantly public element upon it. For that reason the public welfare must be served with adequate state regulation and monitoring of relevant pest control efforts, whether these are publicly or privately exerted (see Chapter 2).

The mosquito control districts of the United States appear to be in accord as to the role of insecticides now and in the near future. The organic insecticides will be useful everywhere into the foreseeable future, and they will be indispensable in some districts (see Chapter 2). As public health pest control continues to develop, pesticides with special attributes will be identified as essential in important but limited pest control programs. The size of the market for such pesticides precludes their commercial development because it is not profitable to manufacture and distribute them.

The study team recommends that some form of federal subsidy be considered to ensure the availability of critical limited-use pesticides at a reasonable cost.

The special cases of DDT and Paris green for public health use are examples of such materials. Although resistance and other problems limit the effectiveness of DDT in many areas, it remains the insecticide of choice in most malarious areas.

Steps should be taken to ensure a continuing and adequate supply of high quality DDT formulations to meet the worldwide demand for domiciliar residual application in public health programs.

Paris green is a chemical which is relatively easy to manufacture, has a unique and effective use as a mosquito larvicide, and when used properly for this purpose, is environmentally safe (see Chapter 2).

The necessary steps should be taken to ensure the availability of Paris green at a reasonable price for mosquito control.

The reservation of certain pesticides for public health use would only rarely, if ever, be feasible economically, and in most instances, because of cross-resistance, it would not achieve the objective of avoiding the development of resistance to pesticides.

The study team does not recommend as a general practice that certain pesticides be reserved exclusively for public health purposes.

To avoid the conflicts that have arisen in some areas, the agricultural and public health authorities should attempt to coordinate the use of pesticides in their areas to serve the best interests of all concerned (see Chapters 2 and 5).

Mosquito control strategies must aim at more than ridding communities of mosquitoes effectively and efficiently. Mosquito control can no longer be a unilateral operation. Collaboration in many directions is now essential--with landowners, with agricultural scientists, with conservationists, with engineers of several sorts, with a great variety of public agencies and, most assuredly, with land-use planners. Mosquito control is now an integral part of land and water management in the public interest (see Chapter 2).

In the future, development and implementation of mosquito control through manipulation of the environment should incorporate appropriate safeguards to preserve the natural vegetation and fauna of aquatic habitats. To this end, more sophisticated forms of environmental manipulation than, for example, simple ditching, should be utilized.

It is difficult to compare needs and results in vector control programs because of the many and varied methods of measuring vector densities used throughout the world.

An effort should be made to develop a universally applicable and acceptable method of population measurement for all the important disease vectors of the world.

Although complete comfort would mean no biting pests at all, complete control at the cost of insecticide resistance or environmental deterioration should yield to reasonable control (see Chapters 2 and 5).

The study team would encourage the establishment of tolerance thresholds for biting nuisances, in the belief that reduction of the pests to lower densities carries the risk of excessive external, if not internal costs.

Managing an organized mosquito control district demands professional leadership. This leadership should not be compromised by lowered personnel standards or by restriction on exercise of professional leadership (see Chapter 2).

Organized mosquito control in the United States has always been and continues to be "integrated control," in the usual sense of the term. The term "alternative methods" is disputable if the implication is that the use of insecticides is all there is to mosquito control. There have always been alternatives, in this sense, and what is happening today is only a reordering of the various chemical and nonchemical methods. The really new "alternatives," such as genetic, biological (other than fish), and hormonal control, are still years from incorporation into the armamentarium of mosquito control.

To assist in making new alternative strategies available to public health pest control, the study team urges that support be expanded in the area of alternative strategies in vector control. Furthermore, there should be increased coordination of efforts in this area because most alternative strategies can be implemented only on a large scale.

Good economic analyses of public health programs are almost nonexistent for two interacting reasons: the general failure of the public health specialists to directly assess trade-off questions and the failure of economists to deal adequately with such issues as the value of health as opposed to illness, the value of human life, and the value of simply having fewer pest mosquitoes about. In addition, there continues to be great uncertainty regarding the technical feasibility of various kinds of control mechanisms. Even unit-cost coefficients for such standard vector control devices as source reduction, larviciding, and adulticiding vary greatly by region and will not be readily available without a major new effort in data collection (see Chapter 2).

The study team recommends that economic analyses be made of the many issues in the vector control field.

Only with this information, which will accumulate only through greater interaction among entomologists, public health specialists, and economists, will it be possible to reach any general conclusions about the economic efficiency of pesticide use versus other approaches to vector control.

Statistics on nationwide use of pesticides for public health purposes are difficult to obtain and are seldom precise when they are obtained. Hence, assessment of pesticide use in the public health field must be approached qualitatively and with historical and current trends as basic criteria (see Chapter 2).

The American Mosquito Control Association or some appropriate federal agency (e.g., the U.S. Public Health Service or the Environmental Protection Agency) should establish improved methods of assessing the levels of pesticide use for public health purposes and keep statistical data on such pesticide use and records on other methods of mosquito control.

Pest control activities for public health objectives are certain to have an impact on the environment. Much of the arthropod-borne disease problem occurs in man-made environments, domestic, urban, or rural, where it is pointless to debate the impact of vector control, if for no other reason than that simple sanitation, which can so often be the remedy, is unquestionably an improvement in environment. It is impacts on "natural" environments that are of considerable concern today, especially aquatic environments. Terrestrial environments are seldom disturbed by the control of disease vectors, with the exception of pesticide pollution or the entrance of pesticides into terrestrial life through the food chain (see Chapters 4 and 5). There appears to be no alternative to use of pesticides in the near future.

The study team recommends continuation of the current trend toward a balanced, integrated approach in pest control tactics taking into account the ecology and behavior of the pest species, the epidemiology, and environmental values. The implementation of multitactic control systems will increase research efforts; consequently, the study team recommends continued and expanded support of research on ecologically based control systems for mosquitoes and other arthropods of medical importance.

Mosquito control operations on the tidelands of the Southeast have the potential to damage the valuable shrimp fishery. However, the impact of present-day mosquito control in this area has only a very minor and very localized impact on the shrimp fishery. Shrimp authorities are more concerned about other activities of man on the tidelands (see Chapter 2).

The study team recommends that attempts to protect the tidelands from various destructive activities, such as dredging, filling and pollution, should exempt mosquito control operations carried out with reasonable concern for environmental values and maintenance of the tidelands environment.

Until recently, the introduction of exotic (nonnative) larvivorous fish for the control of mosquitoes was considered altogether unobjectionable. Advances in ecological knowledge, particularly in aquatic systems, are now exposing areas of concern that demand that this form of biological control be practiced with more vigilance. The introduction of exotic larvivorous fish for mosquito control may pose a serious threat to native fish both large and small (see Chapters 2 and 4).

Introduction of fish into new areas should be made with appropriate evaluation of the impact of their introduction on the endemic biota and also of the impact of the introduction of other undesirable organisms through contamination of the intended introduction.

In investigations extending over more than a half a century, a wide variety of biological controls have been studied. A few of these, e.g., the introduction of larvivorous fish, have become effective and established control procedures where mosquitoes' range is limited. A large number of other biological control agents have been identified, and, in a few instances, their characteristics and limitations described, but the full potential of these agents is largely unknown.

The developed countries may indirectly affect the supply of various insecticides for public health use in less developed countries. They may also have a more direct effect on public health programs by conducting vector control and research programs in other countries. In Chapter 3, we examine the activities of the official agencies of the U.S. government that are concerned with vector control and with vector research outside the territorial limits of the country, and we describe the major international programs for vector-borne disease control, taking note of the role the United States plays in these programs.

Pesticides are essential not only in pest control for public health purposes, but also, of course, in food production. Current practices in pesticide use are sometimes unsound from the standpoint of crop protection, human health and safety, or, in some cases, environmental quality. Improvement is needed in the pesticide management process, i.e., the proper selection, procurement, formulation, packaging, shipment, storage, marketing, application, and disposal of pesticides (see Chapter 5).

The study team recommends an agromedical team approach to the problem of poisoning, persistence, and resistance in pesticide management.

With the widespread introduction of technology from the developed northern temperate countries into the tropical areas after World War II, there has been increasing concern over the undesirable features of this transfer of technology. With this in mind, the study team has considered the impact of the introduction of new agricultural technologies on the problems of public health. Because this is such a broad area of interaction, the study team examined the impact of the changing technology of rice production in Asia (particularly Southeast Asia) on public health problems. A summary of this study is presented in Appendix A. Other examples of this type of interaction are referred to in other parts of the report. We also consider the direct (intoxication and mortality from pesticides) and indirect (induction of pesticide resistance in disease vectors) impact of pesticide use on public health. In this case, we use the situation in Central America as our primary example.

The "green revolution" has not caused sweeping changes in the way rice is grown in Asia. Still, the potential is there for far more extensive changes than have already occurred. In the future, these changes will undoubtedly occur at a slower rate than before, because of the factors limiting the spread of green revolution technology.

Where large areas have been newly irrigated for rice production, and where insecticides have been introduced for the control of rice insects, no drastic changes in patterns of disease transmission appear to have occurred. Part of the reason for this is the relatively small number of disease vectors that are associated with rice fields.

Those vectors that do use rice fields as habitats have been identified. There are only a few areas (Java, South China) in Asia where rice-field *Anopheles* are primary malaria vectors, and a few areas where they play a secondary role. The picture for filariasis is less clear, but rice-field mosquitoes appear to be relatively unimportant except on the west coast of Malaysia, and that focus has contracted sharply as a result of anti-*Anopheles* campaigns. The *Culex* vectors of Japanese encephalitis virus are closely associated with rice culture in many parts of Asia, so much so that there is some evidence, from Korea, that insecticide treatment of rice pests may dramatically reduce the incidence of the disease, at least temporarily. The snail hosts of *Schistosoma japonica* have a wide range of habitats, but in at least some areas (Philippines, China) rice culture may play a significant role in the ecology of the disease.

Where rice culture is important, any change in its production technology, or in the extent of its fields, may change the habitats of the disease vectors discussed above. In most of Asia, rice production has not been "modernized" by the introduction of agricultural chemicals, large-scale water management technology, and heavy machinery. There is still a potential for the development of sound, integrated pest control programs, which would take into account both agricultural pests and vectors of disease.

It is clear that, in addition to pesticides, a wide variety of tactics is available for use in the control of public health pests. For some of these the effectiveness and usefulness in control are well established and scientifically validated. The procedures of source reduction through environmental sanitation, the use and management of larvivorous fish and the use of screening and repellants are among such established alternatives to pesticides. Other alternatives are more limited in their current usefulness, and some are in only the earliest stages of evaluation. Some of these show great promise, and each is discussed in the report that follows. Another critical consideration will be how these several tactics will be used together in pest control or, better, in a pest management system. Identifying a new technology or tactic for pest control is not necessarily the same as identifying a functional alternative control. Such a technology must be made a part of a strategy of pest control and fitted into a management system. This adaptation of a technology into a system must include an economic analysis of that technology, including a benefit/cost study of the particular technology and of the modified system as an operational unit (see Chapter 4).

It is the view of the study team that there is in prospect no alternative to the use of pesticides in public health pest control strategy in the near future. Rather, the study team anticipates a gradual return to a more balanced, integrated approach using a variety of pest control tactics as appropriate to the ecology and behavior of the pest species, the epidemiology of the disease, and environmental values. There will be an increase in reliance on source reduction, and the goal will be containment and reduction of disease rather than eradication.

1

INTRODUCTION

PEST CONTROL AND PUBLIC HEALTH

Man is subject to a wide spectrum of vector-transmitted diseases; representative examples are listed in Table 1-1. Certain of these, such as onchocerciasis in Africa, restrict the settlement by man of certain well-watered river valleys because the local people are well aware of the consequent river blindness. Acquisition of a disease such as African trypanosomiasis, if untreated, is a death sentence; resettlement of peoples away from tsetse-infested foci has been the only effective approach to prevention in some regions. In areas of Southeast Asia, hyperendemic malaria has prevented man's use of regions of great fertility for agricultural purposes. Now, as in the past, vector-borne disease remains a major deterrent to human settlement and agricultural development in some geographic areas.

Nevertheless, during the past quarter-century the use of pesticides together with other tactics to control arthropods has had tremendous impact on improving the health and well-being of man throughout the world. Since the appearance of the residual organic insecticides in the early 1940s, many vector-borne diseases, e.g., yellow fever and malaria, have been virtually eliminated in many parts of the world. In fact, today the control of arthropods in the United States is almost entirely concerned with ensuring man's comfort both inside and outside his home rather than in preventing disease (Reeves 1972a).

Why cannot this level of disease elimination prevail in other parts of the world? The answer is difficult and complex and, in some aspects, must be evasive and incomplete. But as this report will suggest, it will be a long time before vector-borne disease can be reduced everywhere to the level prevailing in the United States

TABLE 1-1 Representative Examples of Vector-Borne Diseases and Control Methods

Disease	Vector	Present Control Methods	Other Promising Approaches	Areas Involved
Chagas' disease	Triatomid bugs (*Rhodinus* spp., *Triatoma* spp., *Panstrongylus* spp.)	Residual sprays (gamma-BHC, dieldrin)	Housing rehabilitation to eliminate breeding sites	South and Central America, Mexico
Dengue and dengue haemorrhagic fever	Mosquitoes (*Aedes* spp.)	Larvicides (Abate, malathion), habitat elimination, adulticides	Genetic techniques	Caribbean region, S.E. Asia, Africa, W. Pacific
Encephalitis (American)	Mosquitoes (*Culex* spp., *Aedes* spp., and others)	Urban sanitation (removal of breeding sites), larvicides, adulticides	Water management, larvivorous fish	North and South America
Filariasis (urban, bancroftian)	Mosquitoes (*Culex p. fatigans* and others)	Urban sanitation, larvicides (fenthion, chlorinated hydrocarbons)	ULV fogging, permanent sanitation measures, drug treatment	Tropical Africa, Eastern Mediterranean, S. America, W. Pacific (Polynesia), Asia
Leishmaniasis	Sandflies (*Phlebotomus* spp.)	Residual spraying (chlorinated hydrocarbons)		India, China, Middle East, Africa, Latin America

	(*Anopheles* spp.)	antilarval measures	(*Gambusia*, guppies), alternative residual sprays (propoxur, fenitrothion), source reduction	areas
Onchocerciasis	Black fly (*Simulium* spp.)	Larvicides (DDT, methoxychlor), fogging		West Africa, Latin America
Plague, bubonic	Fleas	Insecticide dusting (esp. DDT), rodenticides	Health education, sanitation	Foci in Americas, Asia, Africa
Schistosomiasis	Freshwater snails	Habitat management (esp. control of man-made water bodies), molluscicides	Competing snail species, fish predators	Africa, Brazil, Puerto Rico, Venezuela, Asia
Tick-borne rickettsioses	Ticks	Repellents, area dusting	Vaccine, antibiotics, body inspection and removal of ticks	Americas, Asia, Africa
Trypanosomiasis (Africa)	Tsetse fly (*Glossina* spp.)	Habitat management (selective clearing), adulticides (dieldrin, DDT)	Sterile males, suppressive drugs	Tropical Africa
Typhus, epidemic	Lice	Body dusting (DDT, gamma-BHC)	Health education, sanitation, personal hygiene, vaccine, antibiotics	E. Europe, Mediterranean, N. and S. Africa, S. America
Yellow fever	Mosquitoes (*Aedes* spp.)	Vaccination, larvicides (Abate, malathion), habitat elimination, adulticides	Genetic techniques	Tropical Americas, w. and Central Africa

SOURCE: WHO Chronicle, Vol. 25, No. 5, May 1971.

and other highly developed countries. This must not be interpreted to mean that efforts to eliminate vector-borne diseases have been a failure. The elimination of malaria from huge areas of the world and its great reduction in other parts has had such great social and economic benefits that it must be considered among the great achievements of this century. By 1972, malaria had been eradicated from areas inhabited by 745 million people, and almost as many were being protected from it in areas where it had been partially eliminated (WHO 1973a).

Why, then, is there increased concern over the threat of vector-borne diseases and the potential return of these scourges (Reeves 1972a,b)? The great ecological and social disruption associated with war is one explanation of the recrudescence of arthropod-borne diseases and other plagues of man. Such diseases have long been part of the history of wars: huge epidemics of louse-borne typhus and relapsing fever, for example, accompanied and followed World War I. On a smaller scale, the same situation has prevailed in recent times, as in the outbreak of yellow fever in Nigeria in 1969, the increase of plague in South Vietnam from 1963 to 1973, and increases of malaria in Pakistan and India in 1973 to 1974.

Other changes are continually occurring in man's environment and social structure that may create or magnify problems of vector-borne disease. In the past decade, progress toward the elimination of disease suffered setbacks in certain difficult geographic areas as a result of such changes. Environmental and social change can also explain some of the recrudescences. The initial successes with malaria control using DDT are not being repeated in all areas. In some of the problem areas, the habits of man himself in relation to those of the vector are responsible for continued transmission (Wright et al. 1972). The tactics of malaria elimination in these instances must be modified on the basis of detailed epidemiological and ecological information.

Problems are also developing with other diseases of man. As an example, the study team examined problems associated with rice production (Appendix A), but equal attention could be paid to the socioeconomic impact of vector-borne disease associated with and produced by the explosive expansion of cities. The great movement of rural inhabitants into the expanding cities of both temperate and tropical areas, when coupled with the explosive population growth, overtaxes the urban sanitation facilities and creates new exposure to disease and its vectors. Bancroftian filariasis is now reaching epidemic proportions because the mosquito vector, *Culex* sp., breeds in the dirty wastewater collections so common in the slums of tropical cities.

In the temperate areas of some developed nations, there is a reverse movement from the towns into the wooded areas. In the United States, movement to the suburbs has increased man's exposure to such woodland vectors of California encephalitis virus and dog heartworm as *Aedes triseriatus*, *A. canadensis*, and *A. infirmatus*. Similarly, in the Soviet Union, forest pioneers are exposed to spring-summer encephalitis and its tick vector, *Ixodes persulcatus*. These situations will occur with the other kinds of encephalitis viruses as this pattern of human settlement continues and as development of irrigation projects favors the vectors of these viruses (Reeves 1972b).

The creation of man-made lakes and their associated irrigation schemes throughout the world is rapidly extending the range of some diseases, e.g., the debilitating schistosomiases (Hughes and Hunter 1973), and may increase the incidence of other vector-borne diseases, such as malaria and several arboviral infections, in the impoundment areas. Infection with urinary schistosomiasis at rates of up to 95 percent have been recorded among fieldworkers on newly irrigated land in Africa (Sturrock 1965, Foster 1967). Devastating epidemics of malaria were recorded following the construction of the Sukkar Barrage in Pakistan and the Mettur Dam in India. Onchocerciasis transmitted by blackflies often plagues dam construction workers and becomes prevalent downstream from the impoundment.

Containers of nondegradable plastic, aluminum, and glass concern the environmentalist and are also of concern to the public health worker, for rainwater collecting in such containers provides optimal breeding conditions for *Aedes aegypti*, the peridomestic mosquito that transmits yellow fever and dengue.

These examples indicate that public health problems produced by vector-borne disease remain unresolved and that changing life styles of man constantly create new problems.

The use of pesticides for agricultural purposes has of itself created some additional problems in the elimination of disease. In a direct way, these pesticides have caused disease and death, especially when they have been used carelessly. It is estimated by WHO that there are approximately 500,000 cases of pesticide intoxication per year on a global basis, with a 1 percent fatality rate (WHO 1973b). Indirectly, the use of pesticides may jeopardize control of disease by inducing resistance in the disease vector. These two problems are considered in detail in later sections of the report, and the critical situation in Central America is considered as a case study. Since pesticides are essential to ensure an adequate supply of food and fiber for an ever-increasing human population, public health specialists will, for the foreseeable future, be faced with the two problems of accidental poisoning and insecticide resistance. Both agricultural and public health interests are involved in these serious problems, and they should be attacked with an agromedical team approach (see Chapter 4).

An attack on the disease vector is not the only means of reducing or eliminating an arthropod-borne disease. Hence, it is useful to examine the importance of other approaches to the control of vector-borne infectious diseases as a part of total disease control. The other approaches are (a) protection by vaccines, (b) prophylaxis by suppressive chemotherapy, and (c) therapy of human cases to minimize opportunities for infection of specific vectors. It should be emphasized that, even if available, these alternatives to vector control are often not practical.

Vaccines have been developed for only a few of the disease entities listed in Table 1-1. Yellow fever vaccine is the outstanding example. There is no vaccine for dengue, which is currently prevalent in the tropical Americas, or for most of the mosquito-transmitted viral encephalitides. The known existence of multiple antigenic types of the pathogenic arboviruses in this group poses problems in the development of vaccines. Vaccines are available for some of the louse- and tick-transmitted rickettsial diseases, i.e., typhus and spotted fever, as well as for plague; but these do not give perfect protection, and their protective effect is of limited duration. No vaccines exist for diseases caused by the parasitic worms, such as filariasis, onchocerciasis, and schistosomiasis. The same is true for the vector-transmitted protozoal diseases, such as Chagas' disease, malaria, African trypanosomiasis, and leishmaniasis.

Without protective vaccines or vector control, reliance must be placed on prophylactic chemotherapy or on treatment after infection has developed. The chemotherapeutic armamentarium is inadequate. There are no specific drugs that will suppress or cure vector-transmitted viral diseases; treatment of yellow fever, dengue, or encephalitis, therefore, is supportive rather than specific. The use of suppressive drugs to prevent malaria has been widely practiced and is of great value. In some of the malarious areas of the world, however, strains of the parasite have developed that are resistant to the commonly used 4-aminoquinoline or biguanide compounds. In parts of South America and of Southeast Asia, this development has lessened the value of suppressive therapy and has compounded the problems of treatment.

While the rickettsial diseases respond to treatment with broad-spectrum antibiotics, logistical and pharmacological problems preclude the use of antibiotics in suppressive or control programs. Prophylactic drugs have been used on a limited scale in control of African trypanosomiasis and filariasis, but problems involving reactions, toxicity, and the logistics of use have not been surmounted. No drug of proven value is now available for the prevention or treatment of Chagas' disease, although there are promising candidates on the horizon. Extensive research on the chemoprophylaxis and chemotherapy of schistosomiasis has yet to produce compounds that are both devoid of toxicity and effective against all types of the disease.

In summary, the majority of the important human vector-borne diseases cannot be prevented by vaccines or by chemotherapy. Of necessity, their control is based on principles of reducing the source of vectors and the contact between vector and man.

SCOPE OF THIS REPORT

The Public Health Study Team was charged with consideration of the following questions:

I. Contemporary Public Health Uses

What is the current pattern of use of various pesticidal chemicals in public health applications? How much of each major compound is used, and against which pests? What is the pattern of production of these compounds, and for each one what is the relative size of agricultural and public health markets? As a corollary, what would be the predicted impact of a shift away from a given agricultural use upon the surviving public health application? To what degree do United States and international agencies support the development of pest control technology and the purchase of materials used in pest control abroad? What are the characteristics of this "agency market," and how is it related to agricultural applications supported by the same or related agencies?

II. Alternative Strategies

What alternatives to present patterns of chemical control are under development or have theoretical potential against each of the major public health pests? To what extent is the development of these inhibited or assisted by national and international rules, regulations, and customs? What are the anticipated costs and benefits of each potential alternative, compared with those for the present technology? In particular, can we identify internal costs that are added to the current price of public health pest control, and can we determine whether there are external costs that should be added as well in order to compute the cost/benefit ratio to the society as a whole?

III. Introduction of New Technologies

By what international arrangements are new public health and agricultural technologies introduced from such "developed" countries as the United States to the "less developed" countries? Have these mechanisms any important impacts upon the general public health problems of countries of the latter kind? For example, many of the so-called "green revolution" cereal grains require higher inputs of all kinds, including pesticides. To what extent has pesticide use actually increased, and what impact has it had on public health problems in recipient countries? Can the Study Team formulate any general conclusions regarding (a) the effect of exporting agricultural technologies to less-developed countries, and (b) the ease or difficulty of transmitting improved public health methods to such countries using existing mechanisms?

IV. Interaction between Agricultural and Public Health Uses of Pesticides

Agricultural use of pesticides may play a major role in the development of genetic resistance in pests of public health importance. How general is this phenomenon? Does the use pattern of contemporary pesticides indicate a substantial regional overlap in the application of the same chemical for agricultural and public health purposes? To what extent do alternative strategies pose similar problems?

This was a formidable assignment, particularly in light of the limited time available; consequently, the study team was able to complete only part of the task. Following the general introduction to the problem presented in this chapter, the report analyzes contemporary public health practices and the use of pesticides in the United States (Chapter 2); arthropods of public health importance as international problems (Chapter 3); alternative tactics and strategies to the use of conventional pesticides (Chapter 4); and the diffusion of more intensive agricultural technologies and their consequences for public health (Chapter 5 and Appendix A). A general summary of findings and a review of the study team's recommendations in these areas appears at the beginning of the report.

REFERENCES

Foster, R. (1967) Schistosomiasis on an irrigated estate in East Africa. Jour. Trop. Med. and Hyg. 70:133-140, 159-68, 185-95.

Hughes, C. C., and J. M. Hunter (1973) The role of technological development in promoting disease in Africa. *In* M. T. Farvar and J. P. Milton, eds., The Careless Technology. Tom Stacey, London, pp. 69-101.

Reeves, W. C. (1972a) Can the war to contain infectious diseases be lost? Amer. Jour. Trop. Med. and Hyg. 21 (3):251-259.

Reeves, W. C. (1972b) Recrudescence of arthropod-borne virus diseases in the Americas. *In* Symposium on Vector Control and the recrudescence of vector-borne diseases. Proceedings, Pan Amer. Health Organization.

Sturrock, R. F. (1965) The development of irrigation and its influence on the transmission of bilharziasis in Tanganyika. Bull. WHO 32:225-36.

WHO (1973a) Malaria eradication and other anti-malaria activities in 1972. WHO chronicle 27:516-524.

WHO (1973b) Safe Use of Pesticides. Twentieth Report of the WHO Expert Committee on Insecticides. WHO Tech. Rep. Series 153. 54 pp.

Wright, J. W., R. F. Fritz, and J. Haworth (1972) Changing concepts of vector control in malaria eradication. Ann. Rev. Entomol. 17:75-102.

2

CONTEMPORARY PUBLIC HEALTH PRACTICES AND THE USE OF PESTICIDES IN THE UNITED STATES

ARTHROPODS AND THE NEED FOR CONTROL

Regarding the control of arthropods of medical importance in the United States, it is important first to recognize the types of activities involved, the types of arthropods controlled, and the reasons for which this control is exercised. Mosquitoes are by far the most important arthropods subject to control for general public health reasons. It is difficult to assess the exact proportion of the efforts involved, but it is probable that 80 to 90 percent of all the control effort expended in the name of public health in this country is directed toward mosquitoes.

The question must be asked: What constitutes public health application? And this involves the added question: What do we mean by "health"? The World Health Organization defines health as the state of complete physical, mental, and social well-being and not merely the absence of disease or infirmity. By this definition, the application of measures to control pest mosquitoes constitutes a valid part of health protection, and many authorities responsible for mosquito control in recent years have pointed out quite validly that they are engaged in public health operations even though they are not directly concerned with the suppression of disease. Anyone who has experienced the hordes of mosquitoes present in some coastal, northern, and upland areas of North America, or who has read the accounts of the early explorers and developers of the United States and Canada, must recognize the importance of mosquito control in the protection of human health and well-being. Also, the economic development of large portions of the United States would be significantly retarded were it not for the control of pest mosquitoes and other biting arthropods. As economic well-being increases, health tends to improve. By the same token, measures to improve human health, including in particular the elimination of such mosquito-borne diseases as malaria and yellow fever, are usually followed by improvements in economic well-being.

The most significant mosquito-borne diseases in the United States have largely disappeared from public consciousness because they have all but disappeared as a cause of human morbidity or mortality. In the early development of the United States and even as late as the early part of the twentieth century, however, malaria, yellow fever, and dengue were significant public health problems in this country. Their disappearance resulted both from efforts directed toward their elimination and as a natural concomitant of a rising standard of living. Much of the public still believes that the absence of mosquito-borne disease in the United States is due to mosquito control measures. Even if the public were not convinced of this, however, public opinion would sustain control measures directed against biting arthropods purely on the basis of human comfort and the need for increasing amounts of land for economic development and recreational purposes.

The Importance of Disease Vectors in the Absence of Disease

A critical examination of the total vector control effort in the United States, as in other highly industrialized countries, reveals that almost all of the activity is devoted to an increase in human comfort rather than to suppression of arthropod-borne disease. This is a worthy goal in itself, and one that generally has the support of the taxpayers who must ultimately approve and pay for these activities. From time to time, disease outbreaks may require vector control measures specifically designed to eliminate transmission, but these are now relatively rare in this country. It is, therefore, worthwhile to explore how this situation came about, whether it is likely to continue, and what might cause changes in the status quo.

Those engaged in the protection of the public, or in the teaching of the material generally described as "tropical medicine," will attest that the great majority of the important arthropod-borne diseases no longer pose an immediate threat to the people of the United States. Medical students no longer regard discussions of these diseases as relevant to their training. Yet, as Reeves (1972) and others have pointed out, we cannot become complacent about the threats posed by these diseases as long as they are active in any part of the world.

Although it has become almost a cliché to discuss the effect of the jet revolution in travel on the potential for disease introduction, this effect cannot be disregarded. In a recent review of the global problems of imported disease, Bruce-Chwatt (1973) discussed the effect of the increased volume and speed of international travel and trade on the possibility of introducing dangerous diseases into areas that are currently free of such problems. He contrasted the speed and volume of international air travel in 1950 and 1970. In 1950, a trip from London to Hong Kong required 5 to 6 days, and the number of passengers on international scheduled airlines in that year was approximately 6 million. By 1970, the London-Hong Kong trip had been reduced to 8 to 10 hours, and the number of passengers had increased to about 70 million. It is clear that these trends are continuing. The speed of air travel at present is such that an infected person may arrive at his destination long before the end of the incubation period and, thus, before any symptoms are apparent. The new approach necessary to the control of movements of disease is reflected in recent changes in the International Health Regulations (WHO 1969). Responsibility for the recognition of imported disease has now essentially passed from the airport or seaport health medical authority to the general practitioner, whose level of awareness of the problem must be raised if the possible consequences are to be avoided.

The publication *Morbidity and Mortality*, issued weekly by the Center for Disease Control, reports repeatedly the introduction of various diseases into the continental United States. This is coupled with a decreased awareness on the part of many physicians of the difficulties presented by certain "tropical" diseases. In this age of rapid travel, it is essential that physicians keep in mind Maegraith's (1963) advice to physicians always to raise the question: "Where have you been and when?"

To understand the probabilities of the reintroduction of significant arthropod-borne disease into the United States, we should understand why some of these diseases have either disappeared from this country or ceased to be major problems. It is instructive in this connection to examine filariasis, malaria, yellow fever, and dengue.

Filariasis

This may be the least understood of the four diseases discussed here. Filariasis established itself in only a single city in the United States -- Charleston, South Carolina--and in only a portion of that city. Presumably, the parasite was introduced by infected Africans brought to Charleston in the slave trade of the colonial period. The status of the disease in Charleston was discussed by Culbertson (1944) who indicated that the focus finally disappeared around World War II, for reasons not clearly understood.

A similar pattern of limited distribution of this filarial parasite has been reported from various urban areas in the warmer portions of the world. In almost no case is it clear why one city has a high prevalence of the filarial worm while similar nearby cities do not. Examination of mosquito population data, where available, indicates that vector density is not the answer. The vector of filariasis in Charleston was undoubtedly *Culex quinquefasciatus*, a species that is still very abundant in most of the cities along the southeastern and Gulf coasts. The available records seem to indicate that *C. quinquefasciatus* is, if anything, more abundant in many of the cities at present than it was in the early part of this century, because it is very much a creature of pollution. It certainly was abundant enough in Houston, Corpus Christi, and Dallas to act as vector of epidemics of St. Louis encephalitis in those cities in the mid-1960s. Thus, it might seem reasonable to consider many of the cities in the southern United States as filariasis-receptive, but we do not know precisely what is required to establish the infection in a given area.

Although human filariasis disappeared from the United States several decades ago, there have been in recent years an increasing number of reports of human infection with filariae of zoonotic origin (Beaver and Orihel 1965). Undoubtedly, much of the recent increase in reports of this phenomenon is due to increased awareness, but it may also be associated with the increase in numbers of domestic animals: Many of the reports deal with *Dirofilaria immitis*, the dog heartworm. Infections with this organism are usually expressed as pulmonary granulomata, including the "coin lesions" of the lung. *D. immitis* is transmitted by a number of mosquito species, depending on locality, and programs based on the control of pest mosquitoes should assist in decreasing the number of human infections. Dirofilariasis in man is usually without important clinical effect, although pulmonary infarct has been reported (Goodman and Gore 1964) and the coin lesions may be mistaken for lung tumors.

Infections of man by subcutaneous filaria, such as *Dipetalonema tenuis*, a raccoon parasite, appear to be more numerous than those by *D. immitis* and probably make up the bulk of the fifty or so cases of zoonotic filarial infection reported from the U.S. Mosquito control programs as usually conducted in the United States may have relatively less effect on the transmission of this parasite, as human infection takes place in more rural areas.

Malaria

The question of why malaria disappeared from the United States in the years between 1930 and 1950 is an intriguing one, and one not susceptible to simple explanations. Many observers believe that this disappearance was due mostly to sustained and conscious human effort in federal, state, or other programs. This belief is not sustained by an examination of the facts. In 1941, Dr. Louis L. Williams, Jr., in a contribution to a symposium on human malaria, stated that the climate of most of the United States, with its short transmission season, made this the only nation in the western hemisphere in which the natural processes of settlement could, and did, automatically remove malaria from a large part of the country. Residual malaria was eliminated, he believed, by the screening of homes.

In 1882, much of the United States east of the Continental Divide, and especially in the more southern states, was malarious, as was a sizeable portion of California. The malarious areas of the country first expanded and then contracted, reaching a very restricted distribution by the late 1920s. Active programs of malaria control in the United States were organized during World War I in connection with extra-cantonment sanitation. The federal government made financial contributions that could not have been borne by the local communities, but with the Depression, malaria reappeared in many areas from which it had all but disappeared. The overall downward trend in incidence of the disease reasserted itself just before and during World War II, accelerated by Public Health Service programs and by the introduction of DDT (Figure 2-1). The upward trend in the late 1960s reflects cases acquired outside the U.S.

In his series of Heath Clark lectures at the University of London in 1953, Paul F. Russell (1955) of the Rockefeller Foundation addressed the question of why malaria disappeared from the United States. He notes that there were no precise statistics on malaria in the United States until about 1923. There seems little doubt that the disease

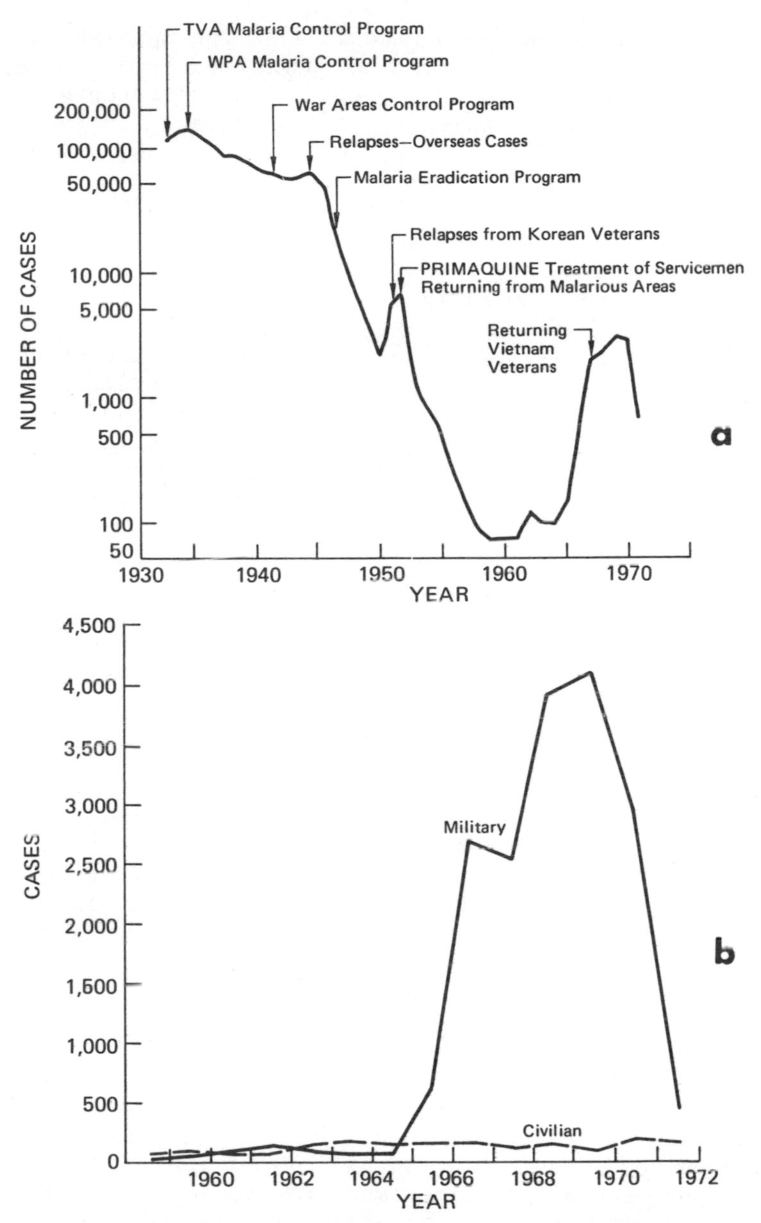

FIGURE 2-1 Malaria. a. Cases by date of report, United States, 1933-1972 (the reported number differs from the count from the case surveillance system). b. Military and civilian cases, United States, 1959-1972, surveillance program.

Source: Morbidity and Mortality, CDC, Annual Supplement, 1972.

was introduced into the New World with the Spanish Conquest. While this is still the subject of some debate, no one questions that, by the end of the nineteenth century, malaria was prevalent in the Southeast, in the upper and lower Mississippi Valley, in eastern Texas, and in parts of California.

> Why malaria has retreated so completely from the United States is a question that has aroused much speculation, especially since there appear to have been no significant changes in the invasive properties of the parasite, vectorial powers of the insect, or immune defenses of the population. Americans overseas readily become infected; paretics and volunteers at home can be inoculated easily with local strains of plasmodia by local anopheles mosquitoes; and the latter will also readily transmit exotic strains. (Russell 1955)

Russell also noted that by 1953 there had been enormous increases in the density of *Anopheles freeborni*, a result of the introduction of rice culture to California in 1912, but there had been no mass reappearance of malaria; only a few sporadic attacks had occurred, including a 1952 episode in which a Marine returned from Korea caused some 34 local cases.

As regards *Anopheles quadrimaculatus*, presumably there has been an overall reduction in numbers since 1875 resulting from the vast amount of land improved for agricultural and real estate purposes, as well as extensive antimosquito drainage since about 1915. Agricultural drainage in the United States has generally been a more effective antimalarial measure than it was in Italy, largely because the American vector *A. quadrimaculatus* has a strong preference for standing water and is not apt to proliferate in drainage ditches as does *A. labranchiae* in Italy.

Russell emphasizes that the mere fact that malaria incidence in the United States began to decline before there were extensive antimalarial projects based on a knowledge of transmission should not be allowed to hide the importance of conscious measures for control during the twentieth century. The agricultural, economic, physical, and social changes have been mostly man-made. In this atomic age, many of these changes could be suddenly reversed. There might even follow a natural reappearance of endemic or even epidemic malaria. Russell concludes that the regression of malaria in the United States has been due to no single cause, but rather to many interlocking factors. Among these, one should not underestimate the effects of conscious efforts at malaria control since 1916.

As noted above, the last sizeable outbreak of mosquito-transmitted malaria in the United States occurred in California in 1952. Since then, there have been a few episodes in which one can only conclude that mosquito transmission took place. However, none of these have caused additional cases beyond the initial brief episode. Nevertheless, there continues to be a fairly large number of reports of malaria cases, often with fatal outcome, in individuals who have been infected outside the U.S. The Center for Disease Control (1973b) reported on one case in Massachusetts that was contracted in Kenya and one in Georgia that was contracted in Zaire, both with fatal outcomes. Between 1963 and 1972, 42 fatal cases of malaria were reported in the United States, 37 caused by *P. falciparum*, illustrating that *falciparum* malaria can take an acute and fulminating course and can be seen at any time in nonmalarious areas such as the United States. Diagnosis of the disease requires a high index of suspicion, which is becoming less common as parasitology is gradually removed from the curricula of medical schools in the United States and as the number of physicians with active military service in malarious areas also begins to decrease.

Examination of light trap and other records from a variety of sources, including articles published in *Mosquito News*, seems to indicate that in the 1960s and 1970s, *Anopheles quadrimaculatus* and *A. freeborni* populations were still at high enough levels to sustain malaria transmission in much of this country. Admittedly, this is a rather subjective assessment, not susceptible of proof in the absence of such transmission. With the large number of military personnel returning from Southeast Asia in 1965-1972 after experiencing malaria attacks, one might have expected some serious outbreaks. That such outbreaks fail to occur seems to be related to the relative isolation of man from the mosquito in present life styles and the probable effect of chemotherapy--but one cannot be sure of the precise reasons.

Yellow Fever and Dengue

These two diseases can be considered together, since they disappeared from the United States at roughly the same time and they share a single vector, *Aedes aegypti*. It is well known that yellow fever epidemics occurred in colonial America as far north as Philadelphia and that terrible epidemics struck New Orleans until 1905.

Dengue appears to have been common as late as 1945 along the Gulf Coast and was the object of *A. aegypti* control campaigns. These campaigns were limited, however, and were not aimed at the elimination of the vector mosquito from the United States. From 1964 to 1968, the United States did engage in an eradication campaign; its results, however, were very limited. There remain significant foci of the vector mosquito in urban centers on the southern Atlantic and Gulf Coasts, and relatively little is being done at any level of government for effective *Aedes aegypti* control in the United States today. The species has reappeared in Galveston after an absence of some years and is known to be present in significant numbers in many other cities--perhaps in numbers great enough to sustain dengue epidemics, but we do not know what numbers are actually required for this.

The species has undergone unexplained population fluctuations in the past, in the absence of control efforts. Hayes and Ritter (1966) reported that *A. aegypti* had practically disappeared from Louisiana by 1966, despite the fact that a significant population had been found in nearby Texas. The population decline began sometime between 1945 and 1957, without discernible cause. Similar accounts appear many times in the literature. In Houston, the species can be found regularly in one sector of the city, but not in others, for no apparent reason. Certainly, the abundance of *Aedes aegypti* is closely tied to the availability of discarded tires, tin cans, and the other artificial containers it requires for oviposition. It may be assumed that dengue has been introduced many times to receptive southern cities since the last outbreaks were reported in the 1940s. In fact, there is a report of a case detected in Florida and acquired in Haiti in an issue of *Morbidity and Mortality* (CDC 1973a). The last case of dengue transmitted in Florida was recorded in 1934.

There are large numbers of people susceptible to dengue throughout the southern United States; there are undoubtedly repeated introductions of the virus from endemic areas in tropical America; and the vector is present in much of the area. What prevents outbreaks from occurring? The most reasonable explanation appears to be that the mode of living of the local population in recent years and the use of household pesticides has diminished the man-mosquito contact below the threshold necessary to sustain epidemics. Unfortunately, we do not know the numbers of mosquito-man contacts necessary for transmission for the United States or for any other area in which the disease occurs.

Yellow fever, with its high fatality rate and obvious pathology, is readily recognized from early American records. The *Aedes aegypti* populations that carried the disease in Philadelphia and other coastal cities were doubtless brought to the United States in the water casks of sailing vessels and found ample containers in the cities for growing populations. It is doubtful that the disease could occur in the more northern cities again, and *A. aegypti* is no longer found as far north as previously. However, much of the southern United States must be regarded as a yellow-fever-receptive area. Unlike dengue, however, surveillance against yellow fever is maintained at a high level, and an effective vaccine is available for the protection of travelers in areas where the disease is endemic. Various indices of *Aedes aegypti* population have been developed, but, as with dengue, we do not have a really good idea of precisely how many man-mosquito contacts are required to sustain an epidemic. It seems likely, however, that of the two diseases, dengue would have the greater chance of becoming epidemic in the United States today.

Other Diseases

In addition to the diseases considered above, there are a number of other diseases in several categories that pose different sorts of problems. The louse-borne diseases, typhus and relapsing fever, and the flea-borne plague (see below) do not appear to pose a large threat at the moment, unless a national catastrophe of major dimensions were to occur first.

The arthropod-borne encephalitis viruses, eastern equine encephalitis (EEE), western equine encephalitis (WEE), and St. Louis encephalitis (SLE) have their natural enzootic cycles so firmly embedded in the ecology of the United States that we will have to deal with them periodically for the forseeable future. With the exception of the situation in California described by Reeves (1972) and the several urban outbreaks of SLE in Texas and the Midwest, the epidemics produced by these viral agents appear to be reasonably limited in time and space. The same appears to be true of California encephalitis (CE), which has come to prominence in recent years. Even where the number of cases is small, public reaction to these diseases is such that immediate response is required. Once these agents are detected in a community, the response, as exemplified by that in the SLE epidemics in Houston, Dallas, and Corpus

Christi in the 1960s, can be rapid and effective. It depends largely on the use of aircraft delivering ultra-low-volume organophosphorous insecticides to eliminate the infected adult mosquito populations as rapidly as possible.

The basic cycle of EEE, WEE, SLE, and CE operates constantly in the United States, and there appears to be no chance of eradicating these diseases. The appropriate measures of surveillance, rapid reporting, and rapid response through the application of immediate mosquito knock-down measures appear to offer considerable promise for the future, as long as the health surveillance and response capabilities remain efficient.

Venezuelan equine encephalitis (VEE) differs from the other agents described above in some respects. The epidemic form does not appear to be present in the United States, although there are foci of what appears to be an enzootic form of the virus cycle in limited areas of Florida. The VEE epidemic that spread into the United States in 1971 was an extension of an epizootic wave that had been moving steadily northward through Central America and Mexico in the preceding year or two. It reached into the United States in relatively limited areas adjacent to Mexico in 1971 and was curtailed by the same tools of surveillance combined with rapid response with mosquito-killing measures that were responsible for interruption of the St. Louis epidemics. Two of the elements that make VEE potentially somewhat more dangerous than the other diseases are its ability to affect many mosquito species with considerable efficiency and the fact that large animals such as horses (and presumably man) may serve as amplifying hosts for the virus.

The various tick-borne diseases like the mosquito-borne encephalitides, exist for the most part in silent foci in the United States and do not appear susceptible to total eradication in the foreseeable future. There have been a considerable number of cases of at least one of these (Rocky Mountain spotted fever) in the United States in recent years but not in such concentrated form as to be described as epidemic. Thus, it seems unlikely that these diseases will erupt in epidemics in the future. Control measures are rarely directed against the tick-borne rickettsiae and almost never against the tick-borne viruses (Colorado tick fever) or spirochetes (*Borellia recurrentis*), which is restricted to caves, summer vacation cabins in the mountains of the western United States, and other such isolated sites.

It is obvious that the United States at present is receptive to several important arthropod-borne diseases.

The primary reason that we are not seeing outbreaks of at least some of these diseases is not so much the existence of any categorical control programs as it is the general economic well-being and the highly organized medical system of the country. It is unlikely that this situation will change very much unless there are major disturbances in the medical system resulting from underlying economic or natural catastrophes of significant magnitude. At present, there is a highly developed reporting system with rapid communication among the health departments of cities, counties, and states and extensive surveillance under the general supervision of the Center for Disease Control (CDC) of the United States Public Health Service (USPHS).

Surveillance

USPHS organized the collection of mortality data beginning in 1878, under the authority of Congress and in connection with quarantine measures against such diseases as cholera, smallpox, plague, and yellow fever. By 1893, a congressional act provided for the collection each week, from states and municipalities throughout the United States, of reports of numbers of cases of specific diseases. Gradually, since the turn of the century, these activities have been strengthened and extended. By 1967, both internal and foreign quarantine programs of the USPHS had been transferred to the CDC in Atlanta.

For the most part, the methods outlined in CDC's *Manual of Procedures* are followed by all responsible health authorities; on the whole, this ensures that disease events of potential public health importance are rapidly reported to the CDC. In addition, the Epidemiological Intelligence Service (EIS), comprised of officers of the USPHS, has been built up. These officers operate many of the central surveillance functions of the CDC and also provide services in a number of areas both in the United States and overseas. Approximately 40 EIS officers are assigned to local and state health departments, and a few act as state epidemiologists. The system functions smoothly, although some of the larger states prefer to handle most of the investigative functions themselves without direct participation by the CDC. Seven hundred EIS officers have been trained over the years, and although many of them have left federal service, they provide an excellent reserve force and act as a source of information in their communities.

If there is any flaw in the total surveillance system, it may be the generally low level of awareness of private and public practicing physicians of the need for alertness in detecting disease of exotic origin. An additional potential problem is the possible loss of the EIS officers, as relatively fewer young physicians are entering the USPHS since the end of the physician draft.

At the time of an outbreak of arthropod-borne disease, such as St. Louis encephalitis, the states and federal government can usually move rapidly to confirm the existence of the outbreak and, with the cooperation of the armed forces and other government agencies, can conduct mosquito control activities rapidly over large areas. It is believed that the aerial spray operations conducted over Houston in 1964 and Dallas and Corpus Christi in 1966, using malathion applied by the ultra-low-volume method, contributed significantly to the reduction in transmission of the virus. Since there were no untreated control areas, it is not possible to state unequivocally that the epidemics would not have ceased without treatment, but the weight of evidence appears to support the contention that aerial spraying and other measures contributed significantly to the cessation of the epidemics.

The situation with regard to control of the epizootic and epidemic of VEE in 1972 is not quite so clear-cut. Because the disease was more prevalent in horses than in humans, the U.S. Department of Agriculture (USDA), as well as the CDC, was involved in the control effort, both in vaccination of horses and in coordination of mosquito control efforts. There appeared to be little coordination among the various federal and state agencies involved in the problem, a deficiency which was particularly criticized by some investigators.

It would appear that there is a need for a permanent coordinating body that would be alerted in case of such epidemics or epizootics and that would bring federal, state, and local agencies together in the most efficient manner. One of the principal officers of the CDC in Atlanta has been assigned the function of coordinating the USPHS aerial spraying, particularly that requested by the Department of Defense. There appears to be a need, however, for a permanent standing committee of various federal agencies to ensure the most rapid, effective, and coordinated response to any outbreak of important arthropod-borne diseases such as VEE.

Conclusions

With the disappearance of such diseases as malaria, yellow fever, and typhus from the United States, we have produced a population almost totally without immunity to these diseases, without necessarily having reduced the potential vector populations. The current life style, economic standards, and general well-being of the American public are such that it seems unlikely that major epidemics of arthropod-borne disease could occur or be long sustained except under most unusual circumstances.

The system of disease surveillance described in the American Public Health Association publication *The Control of Communicable Diseases in Man* (Benenson 1970) should ensure that no serious outbreak of arthropod-borne disease could go without attention for any length of time. Moreover, once an arthropod-borne disease outbreak is detected, the forces and apparatus at the disposal of the local, state, and federal authorities, including ground and aerial pesticide dispersal apparatus, should ensure rapid control.

At present, the economic indicators for the United States appear to be pointing to a period of reduced federal income and expenditure, which will be reflected in equal or larger changes in local funding for public health programs. It seems impossible, however, to gauge the extent of these perturbations in the economy in the next few years and even more difficult to estimate what budgetary cuts may mean in terms of reduced surveillance and reaction capacity in the face of epidemics. It seems reasonable to believe that only a major catastrophe could lead to outbreaks of such diseases as malaria or plague in the United States in the foreseeable future. However, it must be emphasized that all of the elements required to produce an epidemic of several arthropod-borne diseases may be present in this country at any time.

In view of this, it would be well to consider the seven questions raised by Dr. W. C. Reeves (1972) in a Presidential address to the American Society of Tropical Medicine and Hygiene. These questions concern the possible reactions to an outbreak of a serious disease in the United States and the possible restraints on such reactions:

1. When an epidemic occurs that requires the large scale use of an insecticide, vaccine or drug that has not been licensed and it has been declared illegal to use the material, who will make the decision to use it, how quickly, and what action group will arise to oppose the decision?
2. When an epidemic occurs and it is found that the vector or pathogenic agent is resistant to the usual insecticide, antibiotic or vaccine that is available, who will be blamed for not knowing this

has happened and having an alternative material developed, evaluated and available?

3. When the supply of a necessary vaccine, antibiotic or insecticide is inadequate to protect all of an exposed population, who is at fault and who will decide who is to get the benefit of the limited supply?

4. With increased difficulties in financing the cost for development and evaluation of biological and chemical agents for the control of infectious diseases, who is to assume the developmental costs and legal responsibility now that industry is increasingly unwilling to risk the investment or responsibility?

5. When another country or international agency asks us for aid to combat an epidemic, what happens and who takes over, when we say "Sorry, but our economy, foreign policy, supply of materials, knowhow or the social attitude of a significant part of our population will not permit us to assist you?"

6. Who is to be held responsible when decreased priority for infectious disease research or control programs nationally and internationally result in a scaling down of our first line of defense, namely, our research establishment, health agencies, and diagnostic laboratories, and a serious infectious disease epidemic results that is not quickly controlled and emergency funds are not made available for its control?

7. Finally, how long can our society afford to finance vaccination or other widespread control efforts regardless of the recipients' capacity to pay for it or without regard to the dependence of our economy and society on the control effort?

These are not easy questions to answer, and if the possibility of severe economic and social disruption is added to these considerations, along with increasing insecticide resistance, it is not too difficult to imagine that we might once again see extensive arthropod-borne disease outbreaks in the United States, unlikely though this seems at present.

Blackflies and Other Biting Diptera

Blackflies (Simuliidae) are a serious problem in some parts of the United States, not from the disease transmission viewpoint, but as extremely painful and persistent biters. There have been, however, relatively few efforts to control the flies in any systematic manner. Jamnback (1973) has reviewed this field and it appears that within the United States at present the only significant blackfly control program is that in upper New York State.

The most efficient method for blackfly control is treatment with a persistent pesticide of the streams where the larvae occur. The most successful early compound employed was DDT, but since its abandonment methoxychlor and Abate have been found to be excellent substitutes. Approximately 1,000 square miles of New York State, around ten villages, is treated annually at a rate of 0.2 to 0.5 lb of active ingredient per acre in oil solution applied from aircraft. Three such applications are made yearly in the spring, separated by about 2 weeks. Thus, the total amount of pesticide used is quite small, and it is doubtful that any other public health program in the country at present uses amounts of pesticides for blackfly control which approach this.

In the family Ceratopogonidae (the biting midges) there are a number of species, particularly in the genus *Culicoides*, that can cause extreme annoyance, although it is doubtful, as is true of blackflies, that they play any significant role in disease transmission in the United States. Some sporadic control efforts have been directed toward these midges, particularly in salt-marsh and beach areas, and their numbers are undoubtedly reduced during mosquito control activities in coastal areas. Granular insecticide formulations and controlled flooding of breeding sites have been employed against *Culicoides*, but not on a very large scale. Another biting midge, *Leptoconops*, is a problem particularly in California and other western states. It has been controlled by a combination of soil manipulation and insecticide application in very limited areas.

The closely related nonbiting midges in the families Chironomidae and Chaoboridae are extremely annoying and offensive when present in huge swarms and have, at times, been the object of insecticidal control, particularly in California.

Biting flies in several other families (Tabanidae--horse and deer flies; Symphoromyidae--snipe flies; Muscidae--stable flies) may be locally troublesome. The stable flies may be controlled by careful disposal of the moist vegetable matter in which they breed, but for the most part, the most effective measure against these sporadic pests is the use of personal insect repellents.

In summary, the total effort devoted to the control of biting or otherwise annoying diptera is small in the United States at present, and the pesticides devoted to the task are an extremely small fraction of the total public health pesticide usage.

Ticks

The most serious injury to man caused by ticks is the transmission of Rocky Mountain spotted fever. The principal carriers of this disease are the Rocky Mountain wood tick (*Dermacentor andersoni*) in the West and the American dog tick (*D. variabilis*) east of the Rockies. There appear to be no systematic, regular programs for tick control as a means of controlling Rocky Mountain spotted fever (D. E. Sonenshine 1973, personal communication, Old Dominion University). There may be sporadic and localized efforts to control the *Dermacentor* ticks in parts of the Northeast or in the West, but there is certainly no sustained effort comparable to that directed toward mosquito control.

The lone-star tick (*Amblyomma americanum*) has been the subject of much recent and ongoing research. Probably the largest research effort on the control of this tick is centered in the Ozarks and is conducted by Oklahoma State University. Even in this part of the country, however, actual control efforts consist only of sporadic sprayings, mostly of Gardona, in roadside parks and other areas of frequent recreational use (Howell 1973, personal communication, Oklahoma State University). Recreational areas in Georgia, Kentucky, and other states may also receive sporadic sprayings, but no sustained programs now exist.

Fleas and Human Lice

The amount of pesticide expended in the control of human lice and the fleas associated with murine typhus and plague is small when compared to that expended in the control of mosquitoes and filth flies, or even ticks. However, these vectors and the diseases they carry have done much to modify the experience of man on this planet, and their control is still important in many parts of the world.

Human Lice (*Pediculus* and *Pthirus*)

Of the three forms of lice found on man (body louse, head louse, and crab louse), only one, the body louse, is a significant vector of human disease. Body lice serve as vectors for the rickettsial diseases, typhus and trench fever, and the spiral bacterial organisms that cause relapsing fever. Primary among these diseases is typhus, the scourge of armies and perhaps second or third among the great killers in history. With the development of DDT, it appeared that typhus epidemics might be brought to an end, and indeed epidemics were interrupted at Naples and Haifa in the 1940s. However, body lice were among the first insects to show resistance to DDT, subsequently to lindane, and most recently to malathion.

Despite the spreading problem of insecticide resistance in lice, there have been relatively few typhus epidemics since World War II. Perhaps this should not be surprising in view of the general economic well-being of at least the major industrialized nations, since epidemic typhus and relapsing fever are generally associated with malnutrition, poverty, war, and social disturbance.

Louse-borne disease, particularly typhus, however, still occurs in epidemic form in Burundi, Ethiopia, Bolivia, and other nations on several continents. There is a reservoir of the disease in many parts of the world in the form of human carriers. It appears that large numbers of lice are required to sustain an epidemic.

Countries with active louse control programs appear, generally, to be unable to reduce the percentage of infestation below approximately 2 to 5 percent (Pan American Health Organization 1973). It is quite apparent that decreases in body louse infestations go hand-in-hand with increases in the standard of living wherever these elements have been examined, but there is not a major geographical area at present in which lice do not occur, provided there is a reasonably large human population. Precise figures are, however, difficult to come by, and it is probably impossible at present to obtain more than a qualitative estimate of the degree of pediculosis in various parts of the world.

Gratz (1973) has indicated that comparatively accurate information is available on the geographic distribution of foci and outbreaks of louse-borne typhus and epidemic relapsing fever and that body lice are obviously present in such foci; but beyond that, it is difficult to make any positive statements regarding the numbers of lice, as almost no surveys are available outside epidemic areas, and no routine reporting scheme is operating.

It appears (Gratz 1973) that most of the reported increase in pediculosis in the United States refers to head lice and crab lice, rather than body lice; the latter appear to be holding close to the low levels reported for most advanced countries. A number of authors (Gratz 1973, Reeves 1972) have commented on the role sexual promiscuity and communal living may have played in the reported increases of head and crab lice, and similar reports have appeared in medical journals (Ackerman 1968) and the popular press. Quantitative data appear to be available to support the impression of increased incidence of head lice, at least, from England (Gratz 1973).

The extent to which increased louse populations, even if they are the vector body lice, rather than head or crab lice, pose a threat of disease transmission in the United States is difficult to evaluate. For one thing (Wisseman 1973), both an effective vaccine and effective chemotherapeutic agents are available. Also, it must be assumed that lice in the United States are still susceptible to DDT, which has been relatively little used for louse control here, and that the use of DDT would be authorized rapidly in the event of a need to interrupt louse-borne disease epidemics. Wisseman (1973) has emphasized the need for louse control as an adjunct to vaccine and drug use, because the lice represent an immense reservoir of rickettsiae.

The treatment for pediculosis is discussed in Benenson (1970) and in a Pan American Health Organization (1973) monograph. For body lice, powders or dusts appear to be preferable, at a rate of approximately 60 g per adult. DDT is usually used as a 10 percent powder. Lindane may be substituted, either as a 1 percent dust or as a 1 percent ointment. Abate and malathion have also been employed. Estimates of the amounts of these pesticides used for control of lice in the United States have been impossible to derive. Checks were made with health authorities in several southern cities, but no accurate records have been maintained by public health authorities or by dermatologists or other private physicians, who do most of the control work. Pediculosis in itself is not a reportable condition (Benenson 1970), although typhus and relapsing fever are reportable.

Plague, Murine Typhus, Rodents, and Fleas

Although the last major episode of urban plague in the United States occurred in Los Angeles in 1924, sylvatic foci are known to exist in 15 western states, many involving the colonial prairie dog. From 1924 to 1972, 72 cases of human plague occurred in the United States. Only one of these, discussed below, was urban; the others involved contact with rodents or fleas in a sylvan environment. The distribution of plague by county is presented in Figure 2-2 (CDC 1973b).

Plague probably entered the continental United States by ship from Hong Kong at the turn of the century. Some authorities believe that plague has existed in wild rodents in the United States for a much longer period, having come over the Bering land bridge, but there is little evidence to sustain this. At any rate, epizootic plague in wild rodents (*Citellus*, ground squirrels) was detected in California as early as 1903, and human cases traced to ground squirrels were reported by 1908. Since the series of urban plague episodes in San Francisco, Seattle, Tacoma, Los Angeles, and other western cities ended in 1924, almost all transmission of plague to man in the United States has been by contact with wild rodents and lagomorphs or their fleas.

The greatest potential danger from plague in the United States at present is transmission from wild rodents to more peridomestic species and then to the large domestic rat populations in many of our cities. A plague-positive pool of domestic rat fleas was recently found in Tacoma, Washington (CDC 1971). The response was to dust wild and domestic rodent burrows and to increase rodent surveillance.

A preliminary attempt was made to assess the importance of plague in several states, as perceived by the local health authorities, to determine what control measures would be employed, and, if possible, to arrive at some estimate of the pesticides employed for this purpose. Inquiries were made in Texas, where the known foci are quite small, and in New Mexico, where the problem is larger, particularly among some of the Indian groups.

In Texas, a die-off of prairie dogs was noted in July 1973, in the Rita Blanca National Grasslands in the Panhandle (B. Davis 1973, personal communication, Texas State Department of Health). A warning was issued to those who might enter the area, and a check was made immediately to determine if other plague activity was occurring elsewhere in the Panhandle. No such activity was detected. From July 16 to 19, prairie dogs and fleas were collected and sent to the laboratory for examination for plague organisms. Dusting was begun immediately, using 5 percent carbaryl dust, at the rate of 2 oz per prairie dog burrow; in all, 120 lb of the dust was used. The dusting was completed by July 24, the same day that the first prairie dog

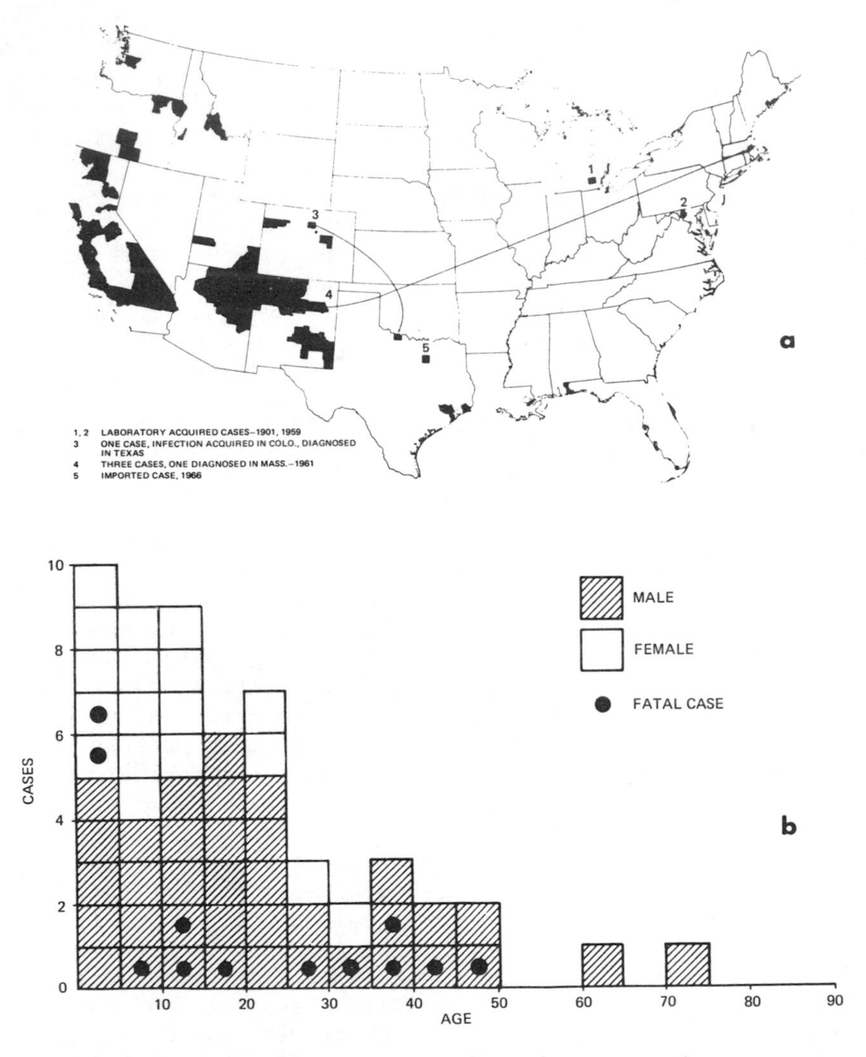

FIGURE 2-2 Human Plague. a. Counties reporting one or more cases, United States, 1900-1972. b. Reported cases by age and sex, United States, 1950-1972.

Source: Morbidity and Mortality, CDC, Annual Supplement, 1972.

carcass was reported positive for *Yersinia pestis*, and before the first positive fleas were reported on August 3. This program required cooperation of state and federal authorities, but rapid coordination permitted completion of protective measures within a week of the first report of the problem.

The Environmental Protection Agency of the State of New Mexico indicated that there had been only a few cases in human beings in that state in the last few years (B. Miller 1973, personal communication) and that no chemical control procedures had been employed. Several applications of carbaryl had, however, been made in prairie dog villages on Indian land in New Mexico and Colorado by the United States Public Health Service, Indian Health Service, in cooperation with the United States Public Health Service Plague Laboratory at Fort Collins.

From all accounts, it appears that an extremely small amount of pesticide is now employed for sylvatic plague control and that no urban plague control is presently under way in the United States. However, as noted above, there appears to be ample opportunity for reintroduction of the vector to commensal rodents in many areas.

A recent and disturbing factor in the plague story is the detection of a focus in Denver, in which the fox squirrel (*Sciurus niger*) was the vertebrate host. This squirrel is a common urban animal in many western and southern cities and might prove a more effective transmitter of the disease to man than many of the other western mammals thus far involved (Hudson et al. 1971).

A further disturbing element is the development of resistance by urban rats (*Rattus norvegicus* and *R. rattus*) to warfarin, one of the most important rodenticides. Neither of these developments poses a serious problem of plague control at the moment, but they require continued attention.

Murine Typhus

This rickettsial disease, which is generally relatively mild in man, causing negligible mortality, is harbored in rats and transmitted to man by the bite of rat fleas, particularly *Xenopsylla cheopis*. At one time, the disease was important in the United States, with approximately 42,000 cases occurring between 1931 and 1946. More recently, there has been a marked reduction in the number of human cases, probably related both to the employment of pesticides for the control of rat ectoparasites and to the introduction of effective antibiotic therapy employing broad-spectrum antibiotics. A number of pesticides have been suggested by CDC and other authorities for the control of rat fleas both for plague and murine typhus control, including DDT, lindane, malathion, diazinon, and carbaryl. With the restriction of DDT use in the United States, the other compounds would appear to be adequate for control of rat fleas, particularly in the case of those infesting domestic rodents. The amounts of pesticides used for this purpose could not be determined for inclusion in this report.

Although there appears to have been a recent increase in the prevalence of human lice in the United States, as mentioned above, this increase seems to have been mostly among the nonvector head and crab lice. Nevertheless, body lice are still present and, along with carriers of epidemic typhus rickettsiae, constitute a continuous if mild threat. The amount of DDT, lindane, and malathion pesticides used annually in louse control could not be determined, but at 60 g per person it probably is insignificant.

Foci of plague in the United States probably will continue to exist for the foreseeable future in sylvan or campestral rodents in the western United States. These pose the threat of transfer to urban rodents, but, barring a breakdown of basic sanitary and public health services, the risk is not great. A small amount of pesticide, mostly the carbamate carbaryl, is used annually in response to reports of die-offs of plague-infested wild mammals or in urban rodent control activities. The amounts so employed could not be determined, but they are probably relatively small.

MOSQUITO CONTROL EFFORTS IN THE UNITED STATES

Mosquito annoyance occurs wherever people reside in the United States and varies only in severity and in seasonality. People do not differ much in their basic tolerance for this annoyance. Beyond a certain level they will insist on control, and below that level whether control is attempted usually depends on how much people know of the possibilities and costs of control and how much they are willing to pay. Small or thinly scattered populations faced with extensive mosquito problems are not likely to have organized control, because the tax base will not be able to support it. The same population dispersed about a large urban area may get good mosquito control simply because protecting the city will necessitate controlling mosquito breeding within a radius of miles, and the tax base will be there to support it. The large metropolitan areas probably all have adequate mosquito control except in their poorest neighborhoods, where *Culex pipiens* populations may exist, along with other symptoms of insanitation.

Where there is no organized public mosquito control, people protect themselves against mosquitoes as best they can, but they generally do not get the information and advice usually available in some form where control is organized. Area-wide mosquito control has come to depend more and more on complicated technology, and the quality of control is dependent on an adequate fund of knowledge. Only when complete and up-to-date information is available and taken advantage of can there be good mosquito control.

Sources of Information

The American Mosquito Control Association (AMCA) is a not-for-profit scientific/educational organization, developed and operated by professional mosquito workers for the benefit of the public, with resources devoted primarily to the extension of knowledge of mosquitoes. The organization's quarterly journal, *Mosquito News*, has a worldwide circulation. Most mosquito control districts have commissioners or supervisory personnel who are members of AMCA and have *Mosquito News* and the AMCA's Miscellaneous Publications on their bookshelves. Many district personnel attend the 3-day annual conventions of the AMCA, where advance information is obtainable on all fronts of the fast-developing mosquito control technology. Next to the AMCA, the most common sources of information are the dozen or so equivalent state or regional organizations, which meet once or twice a year and circulate their proceedings. The New Jersey and California organizations hold annual conventions that attract attendance from well beyond their borders, and the printed proceedings of their conferences are sources of information comparable in many ways to issues of *Mosquito News*.

States with heavy involvement in mosquito control have certain agencies (for example, health department, agriculture department, or university) designated to promote, regulate, or advise in mosquito control matters. These may have entomologists or engineers traveling among the regions or they may issue memoranda and instructional brochures as required by technological, economic, or other developments. Some states conduct schools for mosquito control personnel, as Florida does at its Medical Entomology

Laboratory in Vero Beach. Some prepare operational manuals. In 1973 the California Department of Health's Vector Control Section and the California Mosquito Control Association jointly published a *Training Manual for Personnel of Official Mosquito Control Agencies* (Mulhern 1973). This 224-page, expandable, loose-leaf book, although tailored for use in California, is the only existing modern text on mosquito control, and it is of universal relevancy in philosophy and ideas if not in technical detail.

The USPHS has for decades published outstanding charts and manuals for the identification of mosquitoes and has conducted schools on vector identification, biology, and control. These short courses are sponsored by the National Center for Disease Control either at its Atlanta headquarters or elsewhere, in collaboration with local mosquito control associations. The USDA carries on much research on the technology of mosquito control and publishes results in the journals and proceedings of the associations mentioned above. Recently the USDA terminated its important mosquito identification service when it reduced its taxonomy staff at the U.S. National Museum. The AMCA is trying to carry on this essential service through a well-organized network of taxonomists scattered throughout the country who have volunteered their services.

Representatives of pesticide-formulating firms are a source of information on control technology for all pests, including mosquitoes. Their influence can be good, especially where mosquito control is not well organized. Where official advice on organized mosquito control exists, their sales talk may contradict that advice; but in most cases these representatives offer a valid and essential service. The pesticide industry, by and large, has played a valuable role in the development and implementation of mosquito control.

Unorganized Mosquito Control

How much mosquito control is done on an ad hoc basis by public bodies is not precisely known, but it constitutes a sizable effort nationwide. Many large towns and cities without organized districts suffer mosquito annoyances severe enough for the citizenry to demand action. The result is usually that one or more insecticide sprayings are performed by local public works employees or contracted commercial firms and paid for out of contingency funds. All too often no one involved knows what mosquito species are causing the trouble, or anything about their behavior or ecology, or very much about adulticiding technology. Professional assessment of results is almost never made, and people are satisfied and willing to repeat such control episodes largely because the pests are generally self-limiting populations of *Aedes* or *Psorophora* species whose disappearance can readily enough be attributed to the imposed "control" rather than to nature's. As far as the urban mosquito, *Culex pipiens*, is concerned, there is no question that thousands of city, town, village, and county public health or public works departments are keeping the populations of this species down through ordinary good sanitation practices and enforcement in the absence of organized mosquito control or of much specialized knowledge of control technology.

When mosquito problems are such that they cannot be eliminated by sanitation alone, every state should have available a well-known and easily contacted source of information and advice that any public body can turn to for help. Good mosquito control today is too attainable for the attempt to be left in the hands of the uninformed, especially as toxicants and the environment are involved and because it is so difficult for the public to know whether they have received anything near their tax dollar's worth.

Many public bodies contract with commercial pest control operators rather than conduct the operations themselves. This is no guarantee of better results, although the potential may be there. There still remains the need for an advisory, if not regulatory, body at the state level. The current situation in Houston, Texas, was examined closely in order to shed some light on the sorts of interplay there can be between public and private practice in mosquito control.

Mosquito Control in Houston, Texas

The statutes of the State of Texas (House Bill 127, 1949) permit the organization of mosquito control agencies on a county basis, and set up the system of taxation which supports these agencies at a rate not to exceed 25¢ per $100.00 of valuation. Relatively few of these agencies have been organized thus far, and they are largely restricted to counties near the Gulf of Mexico. In addition to these organized mosquito control districts, a number of municipalities, counties, or other units of government maintain fogging or ultra-low-volume equipment, which are generally used by relatively inexperienced personnel on a schedule in the warmer months or on citizen demand.

In addition, a system has developed around several of the larger cities in which private pest control companies conduct mosquito control operations on a contract basis. Typically, around a large city a real estate developer will purchase a block of land for subdivision. In the absence of zoning ordinances (Houston is the only major city in the United States without a zoning ordinance), the developers generally attempt to control the nature of their developments by a series of restrictive agreements that become part of the deed to the land. The agreements subject each owner to periodic maintenance assessments, and this liability is binding on all subsequent purchasers of the property. The funds so derived are used for such purposes as street lighting, grounds maintenance, and, in most cases, mosquito control. For the first several years of existence of the subdivisions these maintenance funds and the boards that administer them are under the control of the real estate developers, but eventually the boards are taken over by residents on an elective basis.

Some idea of the magnitude of this system is given by the estimate that between 250 and 300 separate subdivisions, such as communities and small groups of townhouses, in the Houston area and nearby portions of Harris County contract for mosquito control services. This is in addition to the Harris County Mosquito Control District, established in 1965 in response to the St. Louis encephalitis virus epidemic of 1964. At the time the district was established it was clearly understood by the political officers of the county and by the appointed members of the board of directors of the district that its mosquito control operations would be directed chiefly against vector mosquito species only (in effect, *Culex quinquefasciatus*), and not against pest mosquitoes. This prohibition was due, at least in part, to the lobbying efforts of the pest control operators (PCOs), who did not want interference with their existing or potential contract business. As a consequence, the Mosquito Control District limits its control activities largely to the agreed-upon areas of the county and its control efforts (fogging) to county parks and other public areas. Initially, the district coordinated its efforts with those of the private operators, at least to ensure that they did not fog the same street on the same night. More recently, this coordination has not been as effective as in earlier years.

Other than this coordination, when it occurs, the governmental bodies have no control or supervision over the PCOs' mosquito control activities--nor are the operators required to report to the city, county, or state the amount of pesticide dispersed, acres covered, or any other aspect of their activities. The only supervision exercised is the rather remote one of requiring the usual applicator's licenses of the PCO and his employees.

A check with the local and state health department and with the Houston Pest Control Operator's Association revealed that none of these organizations had any records of the number of companies engaged in mosquito control, the amounts of pesticides dispersed, or the equipment used. Such operations are known to exist around Houston, Corpus Christi, Brownsville, and other portions of the lower Rio Grande valley. They are not believed to exist around Dallas or Fort Worth--and in Galveston the Mosquito Control District covers the area in such a manner as to make mosquito control unprofitable for commercial operators.

There appear to be approximately six PCO companies in the Houston area that are engaged in mosquito control. Essentially all of this control is adulticiding, with some ditch clearing and oiling by at least one of the companies--the latter on a very small scale. No pesticides (other than fuel oil) are used in larviciding, and malathion is the most frequently used adulticide, with some dibrom used as well. Application is primarily by mist machines, with only one operator presently using ultra-low-volume (ULV) ground equipment. Neither dosages nor costs were determined during interviews with two of the operators.

A decision on how often fogging operations were performed was discussed with one of the operators, the largest in the Houston area. Each year a contract is concluded between the PCO and the community or other unit wishing the service. Based on previous experience (over 20 years) the operator estimates that weekly fogging is required about 28 weeks out of a typical year in Houston. The contracts are drawn on that basis, with the proviso that if fewer applications are required due to decreased rainfall, lower mean temperatures, or other conditions, a *pro rata* reduction will be made for the applications that are skipped. Presumably, no additional charges are made in years in which more applications may be required.

How is "need" determined? Presumably, by customer complaints, as there is no evidence of regular surveillance by mosquito population measurements. Evidently, this method is acceptable to the recipients of the service. The Harris County Mosquito Control District operates light- and chicken-baited traps to provide guidance for their operations and, incidentally, to serve as a source of mosquitoes for virus isolation attempts, but no data on population changes are passed on to the private operators, as far as could be determined. Also, the District exposes caged mosquitoes from time to time to test the efficiency of their ULV equipment, but there is no evidence that the private operators do this. In fact, on at least one occasion a private operator's equipment was found dispersing fog that as far as could be determined by experience and without bioassay, was almost if not totally devoid of pesticide, consisting of fuel oil carrier only.

Overall, these private operatons probably form a small portion of the mosquito control operations, even in Texas. It has not been possible to determine how extensive these operations are in Texas or in other states, but they may be more widespread than might be assumed from cursory examination. Some form of regulation beyond overall licensing may be called for. This problem should be solved in large measure by the application of the certification and record-keeping portions of the amended Federal Insecticide, Fungicide, and Rodenticide Act, which will take effect in late 1976.

Organized Mosquito Control

The majority of Americans living in coastal and other areas subject to an above-average mosquito nuisance problem enjoy the benefits of organized mosquito control. There are approximately 260 organized mosquito control districts in the United States, expending in the neighborhood of $45,000,000 a year in their operations. The analysis that follows is not comprehensive: instead, it relies heavily for specificity in program efforts on 13 representative districts that were contacted for detailed information. The operations of many more districts were considered in the overall assessment, and it is hoped that regional differences in problems and solutions are equitably presented (AMCA 1972, California Mosquito Control Association 1973, Florida Division of Health 1973).

General Dimension and Nature of Programs

The most recent compilation of budgetary statistics for mosquito control in the United States was made by the American Mosquito Control Association for 1971 (AMCA 1972). The data are arranged in Table 2-1 to depict the geographical distribution of the mosquito control effort. The *Southeast*, Virginia to Texas, has the largest concentration of mosquito control effort. The effort is nearly all

TABLE 2-1 Mosquito Control District Budgets as Reported in 1971 to American Mosquito Control Association[a]

Area	Number of Districts Existing	Responding	Total Budgets ($) (responding)	Average Budgets ($) (responding)
Region				
Southeast	97	65	14,683,102	225,893
California	61	53	9,879,474	186,405
North Atlantic	24	24	7,659,266	319,136
Great Lakes	18	11	3,006,702	273,336
West (exc. Calif.)	42	27	1,219,374	45,162
Massachusetts	8	8	1,122,366	140,295
Hawaii	1	1	858,110	858,110
TOTAL	251	189	38,488,394	203,325
States				
Florida	58	48	10,775,022	185,776
California	61	53	9,879,474	186,405
New Jersey	19	19	4,942,032	206,107
Illinois	15	8	1,730,098	216,262
Massachusetts	8	8	1,122,366	140,295
Utah	12	8	499,439	62,430

[a]Eight districts from Mississippi, North Dakota, and Rhode Island, reporting only $3,000 each, were not included.

SOURCE: Data from AMCA 1972.

coastal. Of 97 districts, 65 reported an average expenditure of $225, 893. The Florida districts reporting had an average budget of $185,776. *California* is a region by itself as far as size and diversity of terrain goes. Of its 61 districts, 53 reported annual budgets averaging $186,405. The *North Atlantic* region, with 24 districts reporting an average budget of $319,316, expended the most money per program. New York's 2 districts each budgeted over $1 million, and the 19 New Jersey districts averaged $260,107. In the *Great Lakes* region, the average budget was $273,336 for the 11 districts reporting. Illinois, accounting for 8 of these districts, averaged $216,262 budgets. The *Western* districts have relatively small programs, those reporting averaging $45,162. The budgets of Utah's 8 reporting districts averaged $62,430.

In 1972, five mosquito control districts in the United States had budgets in excess of $1 million, and between 15 and 20 exceeded $500,000. The average budget for the 200 or so best-organized programs is currently about $200,000. Districts vary greatly in what might be called nonoperational expenditures, e.g., administration, insurance, physical plant maintenance, repair of equipment, capital outlay, and reserves. These costs average 30 percent, ranging among districts from 5 to 45 percent in any one year. Also, districts with long nonoperative seasons lend their permanent personnel to other public agencies in off-times so they will not have to hire new crews each year. For these and other reasons, it is nearly impossible to compare districts fairly on overall field control costs. Even in the small sample of districts examined in some detail, areas served ranged from 150 to 2,850 square miles, and populations from 100,000 to 1,846,000. This further illustrates the wide diversity among the districts and the meaninglessness of averages.

The diversity of control programs is further reflected in the mosquito species involved. The 13 districts examined reported no less than 42 species as major nuisances. The house mosquito, *Culex pipiens*, was listed by eight districts, four of which rated it among the top three nuisances. Five other species were listed often enough and with high enough priorities that, with the house mosquito, they constitute the six worst mosquito nuisances in this country: *Culex tarsalis* and *Aedes nigromaculis*, the problem species of western agricultural lands; *Aedes sollicitans* and *Aedes taeniorhynchus*, the salt-marsh mosquitoes of the Atlantic and Gulf coasts; and *Aedes vexans*, the floodplain mosquito of the central and northern states.

The major sources of mosquito annoyance reflect geography, terrain, and degree of urbanization. *Artificial containers and pools*, along with urban and industrial waterworks, reflect degree of urbanization within the district. In most districts these sources account for well under 20 percent of the mosquito annoyance, but in as highly urbanized a district as Orange County, California, they account for 85 percent. *Irrigated lands* are a major source of mosquitoes in the West only, while rice fields are a problem in both West and South. In the Delta Vector Control District of California, irrigated pastures alone are estimated to produce 75 percent of the mosquito nuisance. *Salt marshes* are major mosquito sources along the entire Atlantic and Gulf coasts. In Volusia (Daytona Beach) and

Dade (Miami) counties in Florida and Chatham County (Savannah) in Georgia, they are estimated to account, respectively, for 60, 75, and 80 percent of the mosquito problem. *Temporarily flooded woods, fields, and river floodplains* are the major source of mosquitoes in the North, especially in New England, the Great Lakes states, and the Northwest. *Permanent fresh waters in natural marshes and swamps* are major sources of mosquito annoyance primarily in the interior of the country. They create from 20 to 40 percent of the problem in such urban/suburban districts as New Orleans, Salt Lake City, and Minneapolis-St. Paul (Metropolitan Mosquito District).

Effort Components of Organized Mosquito Control

The work accomplished by an organized mosquito control program can be arbitrarily divided into the following broad components: education, preventive planning, surveillance, environmental manipulation, larviciding, and adulticiding. Some control districts assume other duties, such as research and arbovirus surveillance or control of other arthropod pests, but these are not considered in this review of the common components. As pointed out earlier, it is difficult to assess fairly the costs of these efforts. In what follows, percent of total effort is based on cost as a percentage of total "field" expenditures, i.e., total budget minus nonoperational expenditures. The results are rough but probably as close as one can come to rating these components of effort.

Education In forming a new mosquito control district, it is essential to inform the public so that people will know what they are buying. After the votes are counted and the district is in operation, efforts to keep the public informed often diminish. However, there appears to be a direct relationship between urbanization and the need for continuously educating the public in mosquito prevention and control. The denser the human population the greater the likelihood of domestic mosquito (mostly *Culex* spp.) production from storm drainage and water supply installations, unsanitary disposal of wastes, and neglected water containers. Education is the most effective and least expensive way to eliminate such mosquito breeding.

In highly urban Orange County, California, domestic mosquitoes are the biggest problem but are kept at such low numbers that they are rarely a nuisance. The district maintains an elaborate and obviously very effective educational program, involving the selective distribution of thousands of pamphlets, close work with schools and colleges, the frequent use of all communications media, and the continuing education of personnel at all levels in public agencies liable to create mosquito sources in the performance of their duties.

If there appears to be an inverse relationship between a district's efforts in adulticiding and its efforts in education, there is an explanation. Despite its lower likelihood of inducing insecticide resistance, a large adulticiding program against domestic mosquitoes is regarded as poor mosquito control and justifiable only if there is an active epidemic.

In most districts with large adulticiding programs, mosquitoes originating on wild land are still abundant enough to render domestic mosquitoes of comparatively little concern. Where the area to be protected by insecticides is surrounded by large untreated populations, resistance is unlikely to develop at any considerable rate, because of genetic dilution of the survivors by the surrounding susceptible genotypes. The educational component of control in such districts is straightforward public relations--appearances of the director before civic groups and on television, or the issuance of news releases at opportune times. Such activities usually account for 1 or 2 percent of a district's total budget, but the educational effort in more urbanized districts may consume up to 5 percent and sometimes as much as 10 percent.

Preventive Planning All mosquito control districts do a certain amount of planning to prevent the creation of new mosquito production sources or to eliminate sources already existing. This is a small component of effort, however, in those districts where the major mosquito problem species originate on temporarily flooded wild land, as do floodwater mosquitoes (e.g., *Aedes vexans*), snowmelt mosquitoes (*Aedes stimulans, A. excrucians, A. fitchii,* and many more), salt-marsh mosquitoes (*Aedes sollicitans, A. taeniorhynchus,* and others), and glade and piney-woods mosquitoes (*Psorophora confinnis, P. ciliata,* and others). There, environmental manipulation (discussed below) may reduce or eliminate the problem, but no amount of planning with landowners or public agencies can prevent breeding if rain, thaw, or tide dictate it. With man-made mosquito production, on the contrary, it is theoretically possible to eliminate breeding completely, with the exception of that which occurs in rice fields.

In the case of domestic mosquitoes, collaboration with the right individuals or agencies in the planning stages of waterworks can prevent the creation of breeding sites. This is well illustrated, again, in Orange County, California. The mosquito abatement district maintains constant liaison with the Orange County Flood Control District and the Orange County Water District, both of which in their normal operations continually plan expansions and improvements in water-moving and water-holding installations that could, if improperly designed, create serious and costly mosquito problems. The district also tries to foresee and forestall such flaws in the plans of the U.S. Army Corps of Engineers, military bases, state highway and park departments, and the public works departments of 24 incorporated cities. This preventive planning and the previously described education program are, without a doubt, the most important components of the Orange County Mosquito Abatement District program. They serve well as a model for dealing with the urban problem within any mosquito control district.

Where the mosquito problem is man-made but primarily agricultural, as in California's Central Valley, the possibilities of preventive planning are great. Even if irrigation installations antedated mosquito control, it is still possible to assist growers in modifying their land- or water-use schedules to minimize mosquito production. For many years the valley mosquito abatement districts relied heavily on chemical larviciding, but with the recent escalation of insecticide resistance in *Aedes nigromaculis* and *Culex tarsalis*, the districts are working more closely with growers to prevent mosquito breeding and are enforcing the California laws against creating mosquito nuisances more strictly. Thus, in the Delta Vector Control District virtually the entire work force has turned to preventive work, with even the manager concerned largely with legal citations against noncooperating mosquito producers. It is estimated by the manager that source reduction will account for 90 percent of the district effort.

It is clear from the California experience that the preventive planning component of a mosquito control program can not only be large (e.g., 15 percent and 20 percent in Delta and Orange County districts, respectively) but can also bring about a very considerable reduction of mosquito problems when these originate from domestic-urban situations or from poor agricultural practices.

Surveillance Surveillance is an integral part of every mosquito control district's program. It is the operation that locates mosquito sources, measures and identifies the nuisance, and evaluates the control effort. There are ordinarily three major components of a surveillance program: mapping, inspecting, and processing mosquito collections. Procedure and performance of these functions vary greatly among districts, but certain basic features are common to all.

Every control district needs to be mapped and the mosquito breeding sites identified. How much work this entails varies, but in all districts the labor involved diminishes after the basic mapping is completed. The greater the population growth rate in the area, however, the more effort a district must put into updating its maps. In 1972, even long-established districts spent up to $10,000 on maps and mapping. Maps are indispensable to inspection and to control operations and evaluations and will remain a sizable item in a district's budget.

A thorough inspection program is necessary if a district is to keep constantly abreast of mosquito breeding and aware of the sources and intensity of the problem. All the large mosquito control districts have inspection systems, and these appear the more necessary, paradoxically, the more the original annoyance has been reduced. Mosquito-sampling technology has advanced considerably in the last two decades, so that designing and programming an inspection system has become quite a sophisticated business. There are now very few districts that larvicide or adulticide routinely; most resort to these measures on a basis of need, as indicated by inspection. Light traps and bait traps are widely used as population gauges, but for the precise determination of where to treat, most districts rely on personal inspection for larval and adult mosquitoes. Nearly all the larger districts keep a record of collections and inspections as documentation of accomplishment over the years. All of this presupposes the existence of mosquito identification laboratories, which are an essential adjunct to most large control operations. Even medium-sized districts usually have an entomologist-inspector on their staffs.

Urban areas have special inspection problems because of the complexity of drainage systems, water-supply systems, flood-control installations, and sewerage installations, as well as the disposal problem of artificial water containers. Inspection is often combined with larvicidal treatment, or larviciding may be done routinely for reasons of economy, especially in flood channels and catch-basins. In Orange County, California, and Salt Lake City, the districts spray, respectively, 218 and 122 miles of flood channel. Catch-basin sprayings are routine in most districts, and they can amount to such impressive yearly total numbers of sprayings as 91,148 in South Cook County Mosquito Abatement District (Illinois) and 82,633 in Dade County Mosquito Control Division (Miami, Florida). The Dade County program is also unusual in its long-existing *Aedes aegypti* control effort, which in 1972 involved 59,900 premise inspections.

Because of differences among districts, surveillance as a component of effort is not easily quantified. Among the districts questioned on the matter, it accounted for 13 to 58 percent of the mosquito control effort, and averaged 35 percent. Surveillance was actually the foremost component (among the six) in the majority of programs.

Environmental Manipulation No surer or more permanent control of mosquitoes can be achieved than by so modifying the larval habitat that breeding is prevented. Draining or filling come to mind first, as standing water is thereby eliminated; but the aquatic environment, if natural and desirable, can be retained without producing adult mosquitoes through less drastic yet equally effective manipulations. These methods will be increasingly demanded on noncrop or wild lands as the function of standing waters in ecosystems becomes better understood.

In urban situations, few mosquito control districts themselves correct man-made waterworks that breed mosquitoes; they expect the agency responsible for that installation to correct it. It is clear, as noted above, that preventive planning and cooperation in the design of waterworks is a much better approach to mosquito control than attempting to modify constructions after they are built. The same would be true of agricultural water control schemes, but here the mosquito control director is far more likely to run into a fait accompli and so must develop an improved hydrologic scheme in cooperation with the grower. The actual earthmoving work may then be done by the grower, by the district and partly or wholly charged to the grower, or, if minor, by the district at its own cost. In the West, and particularly in California, the know-how for such corrections is available because of years of collaborative research on the problems. In the Southeast, rice fields, irrigated pastures and groves, and vegetable fields are all, in places, involved in producing mosquitoes, but there has been so little liaison between mosquito control workers and agricultural scientists that solutions are mostly unknown. Much of the resulting mosquito production therefore goes uncorrected or is attacked with chemical larvicides.

Five mosquito problems arising from wild or natural habitats are considered, when possible, for environmental manipulation:

Several mosquito species breed in tree holes, the most noxious of which is *Aedes triseriatus*, a painful biter and an arbovirus vector. Districts with this problem commonly plug the tree holes or treat them chemically. In Minnesota, *A. triseriatus* was found to be the most important vector of California encephalitis. In 1971 the Metropolitan Mosquito District made an intensive effort to find the breeding sites; 585 such tree holes were found, filled with sand or cement, or treated with Abate granules. Mosquito control districts in the Southeast treat the common live-oak tree holes, which also breed that mosquito, in the same manner.

Salt-marsh mosquitoes, the main problem in most districts along the Atlantic and Gulf coasts, can be eliminated by proper ditching or by impounding and flooding the tidelands. These are major manipulations of the environment and as such are not viewed very enthusiastically by ecologists. Nevertheless, all the larger salt-marsh mosquito control districts have resorted to such immobilization of the salt marshes in preference to an endless chemical control program. Salt-marsh water management in New Orleans; Chatham County, Georgia; and Volusia County, Florida consumes, respectively, 27, 33, and 38 percent of the total operational budget.

Floodwater and snowmelt mosquitoes are the major problem in control districts of the northern United States. Although *Aedes vexans* breed in many man-made depressions, the bulk of their production is on wild land bordering streams and lakes. Here, environmental manipulation in the sense of moving earth and water is not readily feasible even if it were permitted in these times of environmental concern.

Freshwater swamp mosquitoes are a small part of the overall problem, but several are key species in the transmission of arbovirus, filaria, and malaria infections. In the past many swamps were drained to control them, but, for ecological reasons, this practice is coming into increasing disrepute. Difficulty of larvicidal control, especially of the widespread *Coquillettidia perturbans*, is leading some districts to experiment with biological control, chiefly the encouragement of native larvivorous minnows or the introduction of exotic ones. So far, however, production of these mosquitoes goes on relatively unabated.

Freshwater marshes are managed in many parts of the country for wildlife production and hunting. Mosquito control districts, particularly in the West, have a long history of cooperating with wildlife managers to minimize production of mosquitoes.

Programs that require the draining and filling of swamps and wetlands will be increasingly constrained by recent laws designed to protect these areas from human modification. These laws are the forerunner of a broader environmental management strategy (reflected in enacted and pending federal and state land use planning legislation) that attempts to classify land according to its ability to support different uses and also attempts to identify areas with high societal value in their natural state that are vulnerable to pressure from human development. The objective of this strategy is to channel development into those areas where the land can carry it with a minimum of social cost and to protect fragile ecosystems.

There is a broad consensus that swamps, tidelands, and wetlands should be classified as areas of critical environmental concern and should be protected from development. Protection may take the form of a withdrawal from all or from some kinds of development. The Delaware Coastal Zone Management Act, which prohibits all heavy industry on the state's coast, is an example of protection by withdrawal. Other states single out wetlands for special treatment and require a permit before land can be modified. Such legislation may require that environmental factors be considered or may go further and create a presumption that development should be prohibited unless it is environmentally compatible or justified by compelling societal needs. California, as a result of a law passed by direct initiative, subjects all development in the coastal band to a special environmental review procedure. In addition, state laws permitting the formation of drainage districts may be reevaluated and local decisions subjected to statewide review. At a minimum, state participation in drainage for insect control as well as other purposes will be subject to environmental assessment procedures where applicable.

Dredge and fill activities carried out by or under the authority of the federal government will also be affected. The U.S. Army Corps of Engineers has jurisdiction over all wetlands covered by navigable waters or washed by the ebb and flow of the tide. The question of which waters are subject to Corps jurisdiction is a complicated, technical one, but in recent years its jurisdiction has been extended to many land uses unrelated to navigation and thus the Corps has gradually obtained authority over much tideland development. Historically, the Corps allowed dredging and filling as long as the navigable capacity of the waterway was not impaired, but this policy is no longer in effect. The Corps evaluates the environmental impact of the activity before issuing a permit and has the authority to deny a permit solely on environmental grounds. In addition, dredging and filling activities undertaken by the Corps are subject to the impact assessment provisions of the National Environmental Policy Act of 1969.

The management of mosquito breeding habitat to augment the natural role of fish as predators on larvae (e.g., minnow-access ditching on salt marsh) is a method of control long used but now receiving more attention. The introduction of exotic fish continues to be very effective and still involves mostly *Gambusia affinis*. However, ecologists have lately criticized the introduction of this aggressive minnow, since it may upset the natural equilibrium among native fishes. Many districts are therefore stocking *Gambusia* only in man-made mosquito breeding sites. Several California mosquito abatement districts are deeply involved in research to improve rice field mosquito control with managed *Gambusia* populations, and one, Butte County, is thoroughly investigating the control potential of annual fishes of the genus *Cynolebias*. In the West, Northwest, and North, stocking of larvivorous fish is a much more common practice than in the Southeast. It is not unusual for districts to stock up to 100,000 or 200,000 minnows a year. The Metropolitan Mosquito Control District (Minneapolis-St. Paul) has an unusual program: in the past 7 years, in cooperation with the state's Department of Natural Resources, it has stocked mosquito breeding areas with 3,513,000 native fathead minnows, *Pimephales promelas*.

Another operation subsumed under habitat manipulation is removal of obstructive vegetation, such as clearing the edges of ditches or areas to be larvicided or removing aquatic weeds that shelter mosquito larvae. This is especially likely to be necessary in the southeastern United States, and the districts there expend up to 10 percent of their total efforts on such work. Chemical herbiciding is also practiced in many districts. In California in 1972, district herbicidal needs were met with a total of 11,128 pounds of toxicants (16 different formulations) and 36,410 gallons of oil.

Environmental manipulation is today one of the most diligently applied mosquito control methods. All the control districts in the country use it to whatever extent their problems, terrains, and budgets will allow. As a component of effort, nationwide, it varies from 0 to 50 percent of total field effort in mosquito control.

Larviciding All mosquito control districts do some larviciding, and it can be a small component of overall effort, e.g., 4 percent in the New Orleans district, or a very large one, e.g., 42 percent in Orange County, California. Before the advent of DDT, oil or oil with pyrethrum added were the only materials used. Between 1946 and about 1966, except for oil in spot larviciding and Paris green on some salt marshes in the Southeast, larviciding everywhere used organic insecticide formulations. Since 1966, largely as a reaction to the development of multiple resistance to insecticides, control districts have turned increasingly to oil/surfactant larvicides, which they either formulate themselves or purchase mixed, e.g., Flit MLO. This is best illustrated in California where usage of the two most used toxicants, Baytex and parathion, dropped 61 percent between 1970 and 1972--from 210,811 pounds to 81,731--while use of oil increased 56 percent--from 339,215 gallons to 528,800. It is also noteworthy that many districts in California and throughout the country are reverting to pyrethrum larvicides. Other districts, especially in the North, are turning more and more to organophosphate larvicides such as Abate and Dursban, which are more environmentally acceptable.

Because mosquito production differs from year to year, the quantity or type of larvicide used from one year to the next must be chosen with care. The Metropolitan District in Minnesota in the past 5 years has averaged 82,291 acres treated with larvicides; in 1972, treating 69,279 acres consumed 30 percent of the operating budget. Butte County, in the Sacramento Valley of California, averaged 136,171 acres treated in the past 3 years; in 1972 it larvicided 87,590 acres at a cost of 14 percent of the total control effort. The 13,720 acres larvicided in 1972 in the Salt Lake City district consumed 33 percent of the operating budget. In the salt-marsh mosquito districts, typical larviciding efforts and costs are exemplified in Volusia County, Florida, where a 5-year average of 19,164 acres were treated; in 1972, the 13,404 acres larvicided cost 21 percent of the operating budget. In Chatham County, Georgia, the 1972 larviciding of salt marsh consumed 22 percent of the district's operational budget, and in Ocean County, New Jersey, a 4-year average for larviciding, mostly on salt marsh, was calculated to represent 23 percent of the budget.

It is clear that many mosquito control districts recently had larviciding programs accounting for 30 to 50 percent of their total efforts. The range today is more likely to be 10 to 20 percent. The trend everywhere in the country appears to be a reduction of larviciding effort attributable to (a) the development of resistance to insecticides by mosquitoes, (b) the lack of effective and economical substitutes for the organic insecticides, and (c) the ecological concern over toxicants in the environment. Two factors may soon reverse the trend away from larviciding: the rapid development of environmentally acceptable materials like new petroleum formulations, pyrethroids, and hormone mimic growth inhibitors; and increasing pressure by environmentalists against any form of habitat manipulation, as is now so conspicuous in Florida.

<u>Adulticiding</u> Adulticiding, unlike larviciding, is not a general practice among the country's mosquito control agencies. Most control districts, including some of the largest, do only spot adulticiding as needed to reduce nuisance mosquitoes at large outdoor gatherings or in recreational or other selected areas.

Malathion is today by far the most widely used insecticide against adult mosquitoes. In the West, where little adulticiding is done, propoxur (Baygon) is frequently used. Pyrethrins are used in several districts: for instance, in 1972, 25,000 acres were treated with pyrethrins in Butte County, California, and 94,970 acres in Minnesota's Metropolitan District. In the Southeast, where most of the adulticiding is done, malathion is the most used insecticide, followed by naled.

In the Great Lakes region, where floodwater mosquitoes are the main problem and source reduction is difficult, adulticiding is a considerable operation; in such districts as the Metropolitan in Minnesota and the Des Plaines Valley in Illinois, it exceeds 200,000 acres of treatment per year. In the Metropolitan District, the 1972 adulticiding operation was calculated to consume 25 percent of the operating budget. In the Northeast, adulticiding is not a common procedure. Most control districts in New Jersey use only water management and larviciding against salt-marsh mosquitoes. Among the salt-marsh mosquito control districts of the Southeast, several exceeded 300,000 acres of adulticiding in 1972, and the operation commonly represented 10 to 20 percent of operating budgets. Florida's control districts spent $1,831,295 on adulticiding in 1972, which was 17.3 percent of the entire state expenditure for mosquito control.

In most mosquito control districts of the United States, adulticiding probably accounts for well under 10 percent of operating budgets. Expenditures as high as 20 percent are probably limited to a few southeastern and Great Lakes control districts. Unlike larviciding, it is difficult to detect trends in adulticiding, mainly because its usage is as variable as the tolerance of different communities of different levels of annoyance.

Summary

The categories of effort into which the work of organized mosquito control districts falls are difficult to quantify for comparative purposes because of the diversity of measuring and reporting criteria. Ditching may be reported as lineal feet dug or maintained, cubic yards moved, man-hours of hand ditching, hours of dragline operation, and many other ways. Larviciding can be measured in terms of acres or miles of waterway treated, gallons of oil or pounds of chemical used, hours of aircraft flown, man-hours of labor, and so on. This confusion can be overcome, but only by a sort of educated compromise.

It is clear that the relative emphasis placed on each effort component is primarily a function of the type of mosquito nuisance that is the district's main problem (Table 2-2). The districts scrutinized in some detail fit into the scheme of the table, as do most mosquito control districts in the U.S., according to the best available information. Urbanization everywhere demands that mosquito control rely heavily on education, preventive planning, and surveillance. Mosquito problems created by agricultural mismanagement find their solutions largely in preventive planning and environmental manipulation, with a strong surveillance effort. Reducing salt-marsh mosquito production relies heavily on surveillance and environmental manipulation, with considerable adulticiding to control the insects that escape these treatments. The nondomestic freshwater mosquitoes are fought mainly with

TABLE 2-2 Emphasis (indicated by number of x's) Placed on Mosquito Control Components as Determined by Type of Mosquito Nuisance

	Component of Effort (explained in text)					
Mosquito Source	Education	Preventive Planning	Surveillance	Environmental Manipulation	Larviciding	Adulticiding
Domestic	xxxx	xxxx	xxx	xx	xx	
Agricultural	xx	xxxx	xxx	xxx	xx	x
Salt marsh			xxxx	xxxx	xx	xxx
Floodwater			xxx	xx	xxxx	xxx
Freshwater marsh			xxx	xx	xxx	xx
Freshwater swamp			xx	x	x	x

larvicides and strong surveillance to guide the effort, while adulticiding is still required to complete the campaign.

It is clear from this survey that organized mosquito control in the U.S. has always been and continues to be "integrated control," in one sense of the term. Although the past two decades have seen a notable reliance on the organic insecticides developed since DDT, the mosquito control districts never completely abandoned the many other components of control that have existed since the beginning of organized control efforts. The term "alternative methods" is misleading if the implication is that the use of insecticides is all there is to mosquito control. There have always been alternatives, in this sense, and what is happening today is only a reordering of the various chemical and nonchemical methods. The really new "alternatives," such as genetic, biological (other than fish), and hormonal control, are still years from operational use.

As for the role of organic insecticides now and in the near future, the mosquito control districts of the United States appear to be in accord. The organic insecticides will be useful everywhere into the foreseeable future, and they will be indispensable in some but not all districts. Wherever the major mosquito nuisance arises from terrain refractory to environmental manipulation for reasons physical, ecological, economic, or political, mosquitoes will be controlled with insecticides or they will not be controlled at all. The manager of the Metropolitan Mosquito Control District, the late A. W. Buzicky, a past president of the AMCA and an able entomologist highly committed to environmental quality, wrote in a letter dated October 31, 1973:

> I can categorically state at this time that effective mosquito control in this area would *not* be possible without the judicious use of carefully selected insecticides. At the same time I would encourage research by qualified workers on alternative techniques to the point of making them *operationally* feasible."

He speaks for control districts in the northern United States and in Canada, where environmental manipulation has limited applicability against mosquitoes produced on wild and limitless land. The situation is not the same where mosquito nuisances are primarily a product of man's misuse of land and water.

Strategies and Problems of Mosquito Control

Strategy in mosquito control has been defined as "killing mosquitoes at such times and places and in such ways that people get maximum relief for every dollar spent and lose nothing of value in the obtaining of this relief" (Provost 1948). On the basis of such expectations, it is often said that a mosquito control program is only as good as its director. This at once challenges the value of programs without directors, which is the case with most "unorganized" mosquito control. These efforts, as explained earlier, are usually in the charge of men who assume the job temporarily or as a sideline, who are largely unaware of the sources of information on the technology involved, and who cannot be expected to perform professionally no matter how well-intentioned they may be. By contrast, the major mosquito control districts seek experienced men as directors or send selected men away to be trained. The new directors are most often young entomologists or engineers, and some have much to learn from the experienced directors, many of whom may not have had as much formal education initially but who have kept pace with the rapidly advancing and proliferating technology of recent decades.

Mosquito control strategies are limited by the nature of the problem faced and by the adaptability of terrain and budget. But even with these constraints, the well-informed director has a variety of strategies and methodologies to choose from. Faced with a difficult decision, he may explore the most recent developments in the problem area. In states with highly developed programs, he may, after visiting other districts, visit research centers. In New Jersey, California, and Utah, these are on state university campuses or at field experiment sites; in Florida, they are at the Division of Health's research stations at Vero Beach and Panama City, or at the USDA laboratories at Gainesville. In less-favored states, mosquito control districts often undertake research on their own, and most of it is creditable. Only in the four states mentioned, however, is there a strong element of guidance from outside the districts themselves. It is instructive in this respect to examine California and Florida, because they are commonly thought to follow conflicting strategies in mosquito control.

The great similarities and dissimilarities in the California and Florida programs are clearly evident in Table 2-3. In total expenditure of funds and in habitat manipulation as reflected in ditching, the two states are strikingly similar, but California uses far more

TABLE 2-3 Comparison of California and Florida Mosquito Control Programs: Statewide Summary Statistics

Program Ingredient	1970	1971	1972
Locally-raised funds ($)			
Calif.	9,713,372	9,879,474	10,225,437
Fla.	6,862,719	7,589,929	8,196,330
Total budget ($)			
Calif.	9,713,372	9,879,474	10,225,437
Fla.	9,337,719	10,064,929	10,595,677
Ditching, new and maint. (mi)			
Calif.	328.42	376.61	339.34
Fla.	376.70	344.70	232.90
Ditching, earth moved (yd^3)			
Calif.	1,216,781	938,872	1,485,822
Fla.	2,979,385	2,717,695	1,897,193
Larviciding materials			
Oil (gal)			
Calif.	339,216	441,711	528,800
Fla.	1,222,910	1,594,119	1,476,618
Toxicants (lb)			
Calif.	227,140	130,790	98,507
Fla.	53,015	51,319	31,175
Adulticiding toxicants (lb)			
Calif.	5,779	8,661	8,693
Fla.	665,719	899,230	848,440

SOURCE: Data from CMCA 1971-1973, and FDH 1971-1973.

larviciding chemicals, and Florida does much more adulticiding. The differences are therefore in strategy and tactics in the area of pesticide use. A closer examination of the situation reveals historical, geographic, and leadership differences as the most likely explanation.

Organized mosquito control began in 1905 in California and in 1920 in Florida. State laws enabling the formation of taxing mosquito control districts were passed in 1915 in California and 1929 in Florida, but Florida formed an antimosquito association in 1922, 8 years ahead of

California. Before World War II malaria control was of major concern in both states; with regard to nuisance mosquitoes, however, Florida concerned itself largely with salt-marsh mosquitoes and California with irrigation-water mosquitoes in the Central Valley. In 1946, California initiated state subvention of mosquito-abatement districts; Florida did the same in 1949. In both states, the public health departments were put in charge of both research and state aid to districts. California abandoned state aid in the late sixties, and all mosquito control research reverted to the universities, but the health department's Vector Control Section maintained a strong guidance position. In Florida, the health department's Bureau of Entomology has retained and expanded its multiple role of research, guidance, and administration of state aid to the districts ($2,399,347 in 1972, or 29 percent of locally raised moneys).

The major mosquito problems in California and Florida can be broadly contrasted. Florida has about 700,000 acres of salt marsh and mangrove swamp in which *Aedes taeniorhynchus* and *A. sollicitans* breed. These insects have always been a grave threat to the state's tourist industry, which, until very recently, was heavily concentrated in the coastal zone. In the interior of Florida, the glades mosquito, *Psorophora confinnis*, although it commonly breeds in irrigated pastures and citrus groves, breeds most prolifically in the poorly-drained flatwoods and glades, where its abundance once caused serious losses of livestock. In California, freshwater mosquitoes have always been the major problem. Although snowmelts in the mountains have always created floodwater mosquitoes in the Central Valley, the problem has been exacerbated by irrigation, an indispensable agricultural practice that has created millions of acres of excellent cropland. *Aedes nigromaculis* and *A. dorsalis*, joined by *Culex tarsalis* and *Anopheles freeborni*, especially in rice fields, are the greatest mosquito problems.

The primary mosquito problems in these two states differ in two important respects: origins and effects.

Effects. The mosquito problems of California are predominantly man-made. Because their source is readily established, they are more susceptible to prevention and, even if the insects survive preventive measures, they are susceptible to larviciding. The mosquito problems of Florida are not man-made but develop in the wild lands--over millions of acres of salt marsh, mangrove swamp, flatwoods, and glades. Altering this terrain is difficult, costly, and ecologically hazardous. Locating breeding over the vast wild terrain is difficult, so total reliance on larviciding is impossible.

Effects. The mosquito nuisance in Florida is a direct and immediate threat to the state's greatest industries, tourism and retirism. Area-wide adult mosquito control, initially made possible with the advent of DDT, has become essential, because no amount of source reduction or larviciding can yet reduce mosquito populations below the tourist and retiree nuisance threshold. In California's Central Valley, source reduction and larviciding together have succeeded in keeping adult mosquitoes below the annoyance threshold of the resident population. Although once commonly practiced, adulticiding in more recent years has been deemed necessary only in unusual circumstances.

Again overlooking domestic, urban and industrial mosquito control, and source reduction in Florida's salt marshes and California's farmlands, the two states offer vivid contrasts in temporary or chemical control. California's adulticiding program is negligible, but Florida's is very large indeed. Ground fogging has long been the most popular adulticiding method in the state. Until 1972, the statewide total for several years had been close to 2 1/2 million gallons of insecticidal formulation. The drop in 1972 to 1 1/2 million gallons reflects the rapid change to the ultra-low-volume (ULV) technique. Ground-fogging toxicants are roughly 75 percent naled and 25 percent malathion. Ground ULV uses predominantly malathion and some naled. Aerial spraying uses mostly naled, with some malathion. The rapidly increasing aerial ULV has so far used these two toxicants approximately equally.

California's array of pesticides used in larviciding has always been greater than Florida's (Table 2-4). Except for one district's use of Abate, Florida has used only Paris green for well over a decade. The latter is now being phased out as Paris green becomes difficult to obtain. California's use of pesticides in larviciding is decreasing quickly, as resistance in the mosquitoes grows. The use of fenthion (Baytex) and parathion dropped impressively from 1970 to 1972. Used in lesser quantities, malathion and chlorpyrifos (Dursban) have not dropped. The two states obviously differ in more than ratios of larviciding to adulticiding.

Mosquito control districts in California and Florida, like those almost everywhere else, are tax-raising, autonomous units of government. Federal and state laws governing

TABLE 2-4 Comparison of California and Florida Mosquito Control Program: Pesticides Used

Pesticide (lb)	State	1970	1971	1972
Larvicides				
fenthion	Calif.	109,912	71,523	51,332
parathion, ethyl	Calif.	72,603	31,986	18,691
parathion, methyl	Calif.	28,327	6,774	11,709
malathion	Calif.	10,478	12,950	11,267
chlorpyrifos	Calif.	5,128	5,902	4,793
Paris green	Fla.	51,800	39,752	14,885
Abate	Calif.	692	1,655	715
	Fla.	1,215	11,567	16,320
Adulticides				
malathion	Fla.	420,001	596,808	605,835
naled	Fla.	233,407	281,477	223,282
fenthion	Fla.	12,311	20,945	19,323
propoxur	Calif.	5,779	8,661	8,693

SOURCE: Data from CMCA 1971-1973, and FDH 1971-1973.

pesticide usages and water management must be observed, as must laws governing personnel management, equipment usage, and financial accounting. But no state agency dictates what control strategies will be employed nor what permissible pesticides will be used. The director's recommendations are usually considered by the district's board of directors or commission and accepted or rejected. He directs operations accordingly. It is customary, nevertheless, for the director to accept guidance from state leadership.

The basic strategy of mosquito control before the advent of DDT was essentially the same in California and Florida. Prevention of breeding was the first order of business, and killing larvae was second. DDT ushered in a third element, adulticiding. The nature of the problem in each state dictated that California should stress larviciding while Florida stressed adulticiding. What has happened since 1946 is a result of the phenomenon of mosquito resistance to the organic insecticides. As this is, for the most part, simply a matter of genetic selection for resistance, a thoroughgoing larviciding program exerts the necessary selection for tolerance against the entire mosquito population, whereas adulticiding exerts selection pressure only against those segments of mosquito populations that create a biting nuisance in populated areas. This is the crux of the whole matter in comparing results and prospects in California and Florida mosquito control.

Before the first appearance of resistance, W. B. Herms and H. F. Gray exhorted the abatement district directors of California not to forsake first principles and abandon permanent control efforts (Herms 1949), while S. B. Freeborn (1949) reminded them that the 1915 law enabling abatement districts defined their primary function as prevention. But larviciding gave extraordinary control. In retrospect it can be said to have worked too well, especially as it abetted the growing conviction among growers that getting rid of mosquitoes was the abatement district's job, not theirs, and hence they could ignore their role in producing the insects.

By 1954, resistance to all chlorinated hydrocarbon insecticides was so widespread in California that talk of "going back" to source reduction was rife. Gray and R. F. Peters vehemently repeated the advice they had consistently given: source reduction first (Herms, 1949). However, the organophosphate larvicides came along, and the illusion that industry would always come through with a miracle chemical was revitalized. Although resistance to the organophosphates appeared in the late fifties and caused some apprehension, the approaching crisis was not much discussed until 10 years later.

Since 1970 California mosquito control has been in a true state of crisis with resistance to all organophosphates widespread in both *Aedes nigromaculis* and the arbovirus vector *Culex tarsalis*. Today, the districts are turning successfully not only toward source reduction and biological control but also toward the enforcement of their long-neglected laws against remediable mosquito production as well. The crisis in California could have been delayed many years if the districts had listened to the admonitions of Herms, Gray, and Peters, and had not been so insistent on achieving the ultimate mosquito control possible with chemicals rather than the reasonable control more compatible with the long view.

The mosquito control directors of Florida embraced DDT with the same lack of restraint as the California directors. When resistance developed, it was where the most larviciding with DDT had been done. As in California, DDT had worked so well against larval and adult mosquitoes that when it failed people were not so much alarmed as infuriated, and the district directors were in trouble.

In 1951 J. A. Mulrennan and M. W. Provost (Anon. 1951) prepared a policy statement for the State Board of Health that described the state's mosquito problems, apportioned responsibilities for control and proposed a state-subvention program matching locally raised money 75 cents on the dollar, with the proviso that these state funds were to be used only on "permanent" control or elimination of breeding. In 1953 this was implemented by state law. Soon afterward, the State Board of Health advised all the mosquito control districts that organic insecticides should not be used against larvae but against adults only, assuring them that this would stave off resistance. In 1956, A. J. Rogers at the Vero Beach research center developed his unique Paris green granule (see below) which was effective against all mosquitoes. Paris green and oil formulation became the only larvicides used in Florida.

Today, adult mosquitoes, after more than 20 years of extraordinarily intense adulticiding, are just beginning to show organophosphate resistance. As a result of the 1953 legislation, the mosquito-producing tidelands of Florida have been immobilized by dragline canaling or diking and impounding. These and other water-management methods have been so effective that Florida's coastal zone has experienced phenomenal growth and is now populated by millions of people who have no recollection of the gravity of the mosquito problem. The districts are also now under unremitting criticism from environmentalists for the damage done to the natural resources of the tidelands. State environmental agencies are now making it so difficult to maintain installations on tidelands that some districts are reverting to exclusively pesticidal operations, and the State Division of Health is abandoning its restriction of subvention funds to source reduction.

For quite different reasons, mosquito control in California and Florida must henceforth be much more sophisticated than it was when insecticides ruled. Both states are in trouble. California's big problem is mosquito resistance to insecticides. The state is taking the only rational options left to it: enforcement of laws against producing mosquitoes and source reduction through cooperation with agriculturists. It can easily manage the current legitimate concern over environmental quality because so little of its mosquito-producing terrain is natural environment. In Florida, the situation is nearly the reverse. There is no severe problem with insecticide resistance, but everywhere there is a conflict with protectors of the environment. Unlike New Jersey, Florida has never established strong liaison between mosquito control and fish and wildlife agencies, and now the future is fraught with troublesome possibilities.

The California and Florida experiences emphasize that mosquito control strategies must aim at more than ridding communities of mosquitoes effectively and efficiently. Mosquito control can no longer be a unilateral operation. Collaboration is now essential with landowners, agricultural scientists, conservationists, engineers of several sorts, a great variety of public agencies, and, most assuredly, land-use planners. Mosquito control is now an integral part of land and water management in the public interest.

PATTERNS OF PESTICIDE USE

The types and formulations of pesticides used for public health purposes in the United States are the same as those recommended annually by the USPHS in a comprehensive article entitled "Public Health Pesticides," published every spring in the journal *Pest Control* (USPHS 1973). Data on quantities used, however, are difficult to obtain because there is no centralized effort to compile such statistics. Data on use by public agencies, at all levels of government, are kept in agency files and are, for the most part, available on demand only. To obtain and collate these pesticide use data for the whole country under current conditions would be a formidable task. In the private sector the data are probably nonexistent. Since the nearest approximation to nationwide public health use of pesticides is in mosquito control, problems in obtaining use data in that area can be examined as a case in point.

Mosquito control organized into programs or districts probably accounts for the bulk of pesticides used against mosquitoes, yet use data are readily available in published, summary form from only two states, California and Florida (California Mosquito Control Association 1973, Florida Department of Health and Rehabilitation Services 1973). Although these states contain 119 control districts, there are still 140 districts in the other 24 states with organized programs (AMCA 1972). The AMCA is an organization of individuals and is not in as good a position to obtain pesticide use data as state mosquito control associations or official state regulatory agencies might be.

By extrapolation from California and Florida data, one can estimate total usage by the country's 259 districts to have been 2,150,000 lb (actual toxicant) of insecticides in 1972. This is a 54 percent extrapolation. The 1964 usage by 300 North American mosquito control districts, applying a 43 percent extrapolation, was estimated to be 7,069,322 pounds (Anon. 1966). To what extent either figure is accurate and to what extent the indicated decrease in total pesticide use is valid cannot be confirmed. The 1964 data were obtained by the research staff of *Pest Control* in cooperation with the AMCA. In 1971 and 1972, AMCA sent out simple questionnaires, not including pesticide use data. In 1971, 73.7 percent of the districts responded; for 1972 the response was "only fair." Another questionnaire will be circulated in 1974 to obtain pesticide use data. Past performance does not augur well for this new attempt.

Many public bodies, particularly small cities and towns, use pesticides in vector control activities which are part of a much larger function, like "public works." Data on pesticide kinds and quantities used may or may not be kept. Nationwide data could be sought only through a national effort delegating the work to properly placed public workers within each state. The EPA may in the future require that such records be kept; this would allow the collection of information on pesticide use.

Private industry's use of public health pesticides is likewise an unknown quantity. If the private enterprise involvement in mosquito control in Houston (see p. 45) is at all representative of what is happening in several large American cities, the total consumption of pesticides could be considerable. However, our inability to obtain pesticide use data for these Houston operations suggests that nationwide data would be impossible to collect from such sources.

Homeowners' use of pesticides in the control of pests of health import is generally assumed to be great. Exact statistics are probably not obtainable. Even if reliable pesticide sales data for supermarkets, drugstores, and garden-supply stores were available, most formulations purchased are used for nonhealth as well as health pests around the home and yard, and prorating them meaningfully would be impossible.

In summary then, statistics on nationwide use of pesticides for public health purposes are difficult to obtain and are seldom precise when they are gathered. Assessment of pesticide use in the public health field is best approached qualitatively with historical and current trends as basic criteria. Consequently, the study team has been compelled to take this approach.

International Development of Pesticides for Public Health Use

The arsenal for public health consists of far fewer insecticides than that for agriculture. The paramount consideration for public health insecticides is that they be sufficiently nontoxic to mammals to ensure their safety for man and his domestic animals. For international use, they must be inexpensive, and, for modern acceptability, they must have a minimal effect on nontarget organisms in the treated environment. The suggestion, occasionally made, that certain insecticides be reserved exclusively for public health use does not appear to be realistic.

Overseas, DDT still remains the main public-health insecticide, because for the control of malaria in most parts of the world there is no compound so inexpensive and effective, and because it has a remarkable safety record (WHO 1971). For mosquito control in the United States, by 1965 the organophosphorus insecticides had already supplanted the organochlorines by 3 to 1. The main organochlorine used at that time was gamma-HCH, and the main organophosphate was malathion. Today the public-health use of dieldrin internationally has become minimal, and the use of organochlorines in the United States has been essentially discontinued. In India it is becoming apparent that the use of persistant insecticides for public health has its limitations (Sharma 1972). As substitutes for DDT (or gamma-HCH or dieldrin) for antimalarial house spraying, only malathion and propoxur among more than 1,400 compounds examined have passed all seven stages of WHO testing (Wright 1971) and are recommended for use. Fenitrothion is now in the final stage of extensive districtwide testing (Schoof and Taylor 1972). As larvicides for control of culicine mosquitoes, fenthion, chlorpyrifos, and temephos (Abate) are coming into widespread use, while naled and carbaryl have been applied as space-sprays against adults of those species.

All of these compounds were first developed as agricultural insecticides, and their producers depend primarily on the agricultural market. Only the organophosphorus compound temephos (Abate) and, more recently, the juvenile hormone mimic methoprene are essentially specific insecticides for the control of aquatic diptera of public health importance. In the case of Abate, extremely nontoxic for mammals, it appears that the public health market alone is sufficiently large to support industry production. This chemical is now the preferred mosquito larvicide in nearly all situations, not only against anopheline larvae for malaria control (Fontaine and Rosen 1973), but also against *Aedes aegypti*, often in indoor sites. It is also highly effective against larvae of the *Simulium* vectors of onchocerciasis.

Dependence on the agricultural market has in certain cases hindered the development of a public health insecticide arsenal. For example, although chlorpyrifos-methyl is more desirable than chlorpyrifos because of its greater safety for man and aquatic nontarget organisms, only chlorpyrifos, a more effective agricultural insecticide, is likely to be available. The advancement of the promising carbamate Landrin to the final stage of testing for malaria control had been delayed while the producers developed sufficient data on its plant residues to obtain its registration as an agricultural insecticide. Now, however, it is being employed on a trial basis in Central America (Hobbs and Miller 1973). In special public health situations where the development of successive insecticide resistances can be solved only by a specific new compound (for example, a juvenile hormone mimic, which is noncompetitive in all other market situations), the production of that compound could be assured by paying premium prices or by some other form of subsidy.

A number of insecticidal compounds that are related to the DDT molecule but that, like methoxychlor, are biodegradable and do not accumulate in the food chain have been synthesized and tested at the University of Illinois (Metcalf et al. 1971a) and the CSIRO, Australia (Holan 1971). Many of them are more highly larvicidal than DDT to normal as well as DDT-resistant strains of *Culex* mosquitoes (Metcalf 1972). Two of them are as effective as DDT as residual deposits against *Anopheles albimanus*. Their high biodegradability and low biological magnification has been quantitatively measured in a model ecosystem (Metcalf et al. 1971a,b, Kapoor et al. 1973), and some of them are under examination by commercial companies.

The pyrethrins available from dried pyrethrum powder have long been employed to kill body lice and adult flies and mosquitoes indoors; recently, the synthetic pyrethroid allethrin has been widely used in domestic aerosols. The addition of antioxidants, ultraviolet absorbers, and synergists such as piperonyl butoxide have greatly improved the longevity and effectiveness of pyrethrin sprays. A number of new synthetic pyrethroids (e.g., resmethrin, bioethanomethrin) are now coming on the market or are up for registration; several of them may prove to be, like allethrin and dimethrin, effective for control of adult and/or larval mosquitoes (Hayashi and Takama 1972). These would have to prove themselves competitive with the organophosphorus compounds, which are highly effective for adult and larval control except in the areas where resistance has developed, and with certain carbamates (e.g., carbaryl) that are effective for adult abatement. When applied to water, pyrethroids tend to be safe for nontarget aquatic invertebrates except for mayfly nymphs, which are the most sensitive; they are, however, somewhat toxic to fish.

No cases of developed resistance to pyrethroids has yet been recorded for mosquito larvae, although strains already resistant to organochlorines and organophosphorus compounds are somewhat cross-tolerant to them and therefore could be expected to develop resistance on being exposed to pressure from regular use of pyrethroids. Ultra-low-volume sprays with synergized pyrethrum dispersed at 0.004 lb per acre have given excellent control of adult floodwater mosquitoes over 50 square miles in Minnesota. For domestic use, Japan has imported annually some 2,000 metric tons of pyrethrum mosquito coils (Smith 1973a); recently, allethrin has often replaced pyrethrum in the burning mixture (Chadwick 1970). Louse powders containing 0.2 percent pyrethrins plus 2 percent piperonyl butoxide are still recommended but do not give as much residual protection as DDT or malathion powders.

Unrefined petroleum distillate oils have been in use for 80 years as mosquito larvicides, and until recently there has been little change in formulation and dosage. The simplest form is diesel oil plus 0.5 percent linseed oil, applied at 20 to 40 gal per acre; in heavily polluted waters populated by *Culex fatigans*, as much as 80 gal per acre is often required. It is more usual to make a 3:7 mixture of diesel oil with crude oil (heavy fuel oil), and add 0.5 percent Triton-X-100 surfactant and 2 percent creosote as a toxicant; dosages for *Anopheles* larvae are 10 to 20 gal per acre.

An investigation of 110 petroleum-hydrocarbon fractions found 56 that were at least twice as larvicidal as No. 2 diesel oil (Micks et al. 1967). Consequently, an oil was developed, called Flit MLO, that was 35 times as effective as diesel oil. This oil has a low viscosity and a high boiling point and is essentially nonphytotoxic. It is effective against larvae and pupae of *Anopheles* and *Culex* at 1.0 gal per acre. Like three other oils, it did not induce resistance in *C. fatigans* in 20 or more generations of selection pressure (Micks et al. 1968). As it is effective against insecticide-resistant populations, it has been employed against *Aedes nigromaculis* in California; in the same way, the failure of organochlorine larvicides against *C. fatigans* 20 years ago dictated a return to diesel oil. Other highly effective larvicidal oils are now being developed.

Effects of Restrictions on the Use of Pesticides

The removal of DDT from registration for virtually all uses in the major developed countries has so far not resulted in its becoming unavailable for its essential use in antimalaria operations. Governments of developing countries understandably showed some concern about using an insecticide considered so undesirable by those of developed countries. Since 1969, however, environmentalists and medical entomologists have agreed that the antimalarial use of DDT, which is restricted to application inside houses, would have negligible deleterious effect on the wildlife of the countryside; indeed, the quantities of the insecticide that would escape into the open would be of the order of one-thousandth of the dosages applied in agriculture (Brown 1972). In this context, the benefits from this insecticide for human health far outweighs the risks to the environment under conditions of supervised use (Bruce-Chwatt 1971).

The major concern has now become that of ensuring the continuation of the supply of DDT, a product of low profit margin that can be sold at a low price mainly because it is a by-product of more profitable syntheses. At present the need of malaria programs for 45,000 tons annually, more than half the total annual world production of DDT, is largely met by a single supplier in the United States; it is fortunate that this supply, which is of reliable quality and is always delivered on time, has not been restricted. Moreover, a group of African nations has approached UNIDO in Vienna for assistance in establishing production plants in areas where DDT is needed. If DDT were to become unavailable to the malaria program, there could be a resurgence of this disease of the magnitude exemplified by the experience of Sri Lanka (Ceylon), where in 1968 the number of cases rapidly increased from nearly zero to no less than 2 million.

On a smaller scale, several other insecticides virtually essential for public health uses may be in jeopardy, as they are under review by U.S. regulatory agencies on environmental grounds. Gamma-HCH and Paris green (discussed in the next section) are cases in point, as are some aquatic herbicides valuable in mosquito larval control. U.S. agencies that oversee technical assistance to less developed countries are greatly concerned by impending supply shortages of pesticides; many producers have already had to resort to an allocation system. Moreover, there have been almost crippling increases in price. Even that of DDT has almost doubled in the past 10 years. The World Health Organization is now estimating, by means of computer programs, the potential effects, in terms of human deaths, of the lack of pesticides on control of malaria, dengue haemorrhagic fever, trypanosomiasis, and schistosomiasis.

The Status of Paris Green in Mosquito Control

Paris green is a copper-aceto-arsenite of which 56.62 percent is arsenic as AS_2O_3. Along with London purple, it was, in 1860, the first stomach poison used in insect control. Paris green enjoyed great popularity in the control of such chewing insects as Colorado potato beetle and codling moth. After the introduction of lead arsenate in 1892, its use declined steadily. By the time DDT arrived on the scene, Paris green was little used as an insecticide in agriculture.

Arsenic is found in all soils, varying from a fraction of 1 ppm to 40 ppm, and in the natural vegetation it occurs in concentrations up to 10 ppm (Williams and Whetstone 1940). In waters uncontaminated by arsenicals, fish contain up to 5.7 ppm, total dry weight, or up to 60 ppm in body oils (Ellis et al. 1941). When arsenicals are added to soil or water, concentrations of arsenic rise, but they fall again within a few days. What happens is unclear (De Benedetti 1935), but is was agreed long ago that the added arsenic seemed to disappear, and this conclusion was recently affirmed with reference to Paris green used as a submersible mosquito larvicide (Rathburn 1966).

Paris Green and Malaria Control

In 1921, Barber and Hayne described the first use of Paris green to destroy anopheline mosquito larvae. These larvae feed at the water surface, largely by ingesting food particles swept into their gullets by their rapidly rotating mouth brushes. Floating particles of Paris green are similarly swept in. The material, a powder, was mixed with a diluent dust and floated on the surface of breeding waters, where it was ingested by the larvae. In 1926, King and Bradley demonstrated its dissemination by plane.

Paris green was soon being used in enormous quantities worldwide in malaria control, earning an enviable record for effectiveness, economy, and safety (Bishopp 1949). Because of its insolubility in water, it was first thought inappropriate for culicine larvae, which, unlike the anophelines, feed mostly below the surface.

As used in malaria control, Paris green was so safe to vertebrates, cold- or warm-blooded, that it was in many places used against mosquitoes in drinking water for humans. Its record of safety for wildlife was equally good (Bishopp 1940). Among aquatic insects and invertebrates it appeared to have an effect only on filter feeders that ingested a narrow range of particle sizes; this category is virtually restricted to mosquito larvae. As a final tribute to its record in malaria control, Soper (1966) stated that in the only two cases of anopheline eradication--"*Anopheles gambiae* in Brazil in 1940 and in Egypt in 1945--the basic method used in each country was a straightforward chemical attack with Paris green."

With the advent of DDT in World War II, both oil and Paris green larvicides quickly disappeared from most malaria control programs. Never having been satisfactorily developed for use against culicine mosquitoes, Paris green soon disappeared from the scene as a mosquito larvicide.

Early Trials against Culicines

In 1927 Griffitts reported efforts to kill salt-marsh mosquito larvae (*Aedes taeniorhyncus* and *A. sollicitans*) with mixtures of wet sand and Paris green to sink the latter to the bottom. Throughout the thirties there were attempts to formulate Paris green for use against culicine larvae (Herms and Gray 1940). In 1935, King got a good kill of *Aedes taeniorhynchus* larvae by spraying a mixture of Paris green and water, and in the following year, at the request of Claude Strickland, director of mosquito control in Pinellas County, Florida, the U.S. Department of Agriculture did some experiments in that county's salt marshes that looked very promising (King and McNeel 1938). This, however, is as far as anyone ever went toward developing Paris green for use against culicine larvae until the late fifties.

Resurrection of Paris Green

In 1956, Claude Strickland was still director of mosquito control in Pinellas County, Florida, when an enormous brood of salt-marsh *Aedes* hatched there, just outside St. Petersburg. He recommended the use of Paris green because he wished to abide by the State Board of Health's warning to use none of the organic insecticides on larvae. The methodology did not exist to treat 5,000 acres of marsh in this way. The State Board of Health, on the other hand, while agreeing that the situation was an emergency, approved the use of granular parathion instead. However, John Mulrennan, director of the board's Bureau of Entomology, at once requested the Entomological Research Center (now Florida Medical Entomology Laboratory) to investigate the possibilities of Paris green as a salt-marsh mosquito larvicide. In this way, the same man who set in motion the first investigations of Paris green for this purpose 20 years earlier (King and McNeel 1938) played the same role in the reopening of research.

In 1957, A. J. Rogers at the Entomological Research Center in Vero Beach solved the problem of liberating Paris green throughout a water column by using vermiculite as the granular base and sticking the Paris green to the base with a water-miscible sticker (Rogers and Rathburn 1958). The formulation was soon improved (Rogers and Rathburn 1960a) and its suitability for aerial application demonstrated (Rogers and Rathburn 1960b). The long record of safety to man and animals was upheld in all work with this new formulation. In the mid-sixties, the U.S. Department of the Interior gave approval for the use of this Paris green formulation on mosquito breeding waters within its sanctuaries, a singular recommendation. In addition, Rathburn showed that after 31 generations of selection in the laboratory by exposure to LD_{80} dosage levels, *Culex nigripalpus* had developed no significant increase in tolerance to Paris green (Rathburn and Boike 1973).

Loss of Paris Green Larvicide

Through most of the sixties, Rogers Paris green formulation was used in the larger mosquito control districts of the southeastern United States. Although it was effective against all mosquito larvae except the *Mansonia*, its greatest application was against salt-marsh mosquitoes. Many control districts purchased the raw materials, adjusting the formula to suit their own dispersal equipment and field situations.

In the late sixties, the two eastern suppliers of Paris green abandoned its manufacture. The reasons given were little demand (agriculturists no longer used it, and mosquito control directors were swinging over to oil/surfactant formulations that were cheaper to apply), and concern over restraints being placed on arsenical compounds by federal and state environmental and regulatory agencies. One California firm continues to manufacture Paris green, but shipping costs to Florida and the other southeastern states make the price prohibitive, especially in competition with oil/surfactant formulations.

The Future

There are still many situations in which the Rogers Paris green formulation is more effective than currently available combinations of oil and surfactant. The competitive advantage of the latter and of toxicants like Abate and Dursban is largely, though not entirely, economic. Restraints on habitat manipulation, like those in Florida, where large mosquito control districts are selling their earth-moving equipment and returning to the total chemical control they were persuaded to abandon 20 years ago, can only bring about more larviciding. But oil and other petroleum derivatives could conceivably be declared environmentally unacceptable, and mosquitoes could become resistant to the alternative chemical larvicides. This would leave Paris green as a final resort--were it available. If its use were economically and otherwise feasible in other parts of the world, particularly Africa, where the organic insecticides have had scarcely any effect on anophelines, Paris green larvicide could guarantee a longer life to residual applications of such insecticides by forestalling the development of resistance. In any event, Paris green is a simple compound, and there may be merit in encouraging its worldwide production.

In November 1973, the EPA accepted the labeling of Paris green for mosquito control submitted by the sole remaining manufacturer of Paris green in the United States, but with one requirement that could lead that company to stop manufacturing this chemical. EPA wanted them to establish the LC_{50} of Paris green for rainbow trout and bluegill sunfish at 24, 48, and 96 hours. Dr. A. J. Rogers of the Florida Division of Health has furnished data to EPA that, it is hoped, will make the agency withdraw its requirement. The Rogers data reveal that (a) in 1968 a government laboratory (now in EPA but then in the U.S. Bureau of Commercial Fisheries) demonstrated that Paris green as used in mosquito control had no effect on populations of a common salt-marsh fish, *Cyprinodon variegatus* and a grass shrimp, *Palaemonetes pugio;* and that (b) Paris green at dosages up to 15 times those used in mosquito control had no effect on two other common larvivorous minnows, *Mollienesia latipinna* and *Gambusia affinis holbrooki*.

This ongoing campaign to save an effective and ecologically acceptable mosquito larvicide that is inorganic and thus a substitute for the resistance-prone organic insecticides is another example of the results of lack of effective liaison between users of public health pesticides and the EPA. It is also an example of the abandonment of a highly commendable technique for economic reasons. The history of public health pest control is replete with cases of more environmentally acceptable and equally effective technologies being bypassed for other technologies only slightly cheaper.

Rodenticides and Rodent Control

Rodenticides are another type of pesticide generally considered in the realm of public health pest control. In the United States today, three types of rodents are considered pests for various reasons: commensal rodents, field rodents, and structural or stored-product rodents.

Commensal Rodents

Rodents living in proximity to man are the rodents that best fit the definition of a "health" problem. Even these, however, seldom actually transmit disease. Bubonic plague, carried by fleas often associated with rats, still turns up at a rate of one or two cases per year in western rural areas of this country, but it is unlikely to become established in commensal rodents here under present conditions. Murine typhus, also spread by fleas on rats and at one time a significant problem in some southern cities in the United States, has decreased markedly in recent years, particularly since the advent of effective antibiotic therapy. Other diseases that are sometimes associated with rats in this country (such as leptospirosis and salmonellosis) are most often caused by the contamination of food or water. The solution to the disease problem in this case is really one of improving general sanitation rather than merely eliminating the rodents.

Rats are controlled for psychological as well as for health reasons and are the most important of the commensal rodents. A 1967 National Pest Control Association (NPCA) survey found that rodent control by commercial pest control operators is most often directed at Norway rats (43 percent); house mice (39 percent); tree squirrels (5 percent); and roof rats. Rats may bite as many as 60,000 people each year in this country, mostly children and the elderly, and almost always in urban slum areas. Rat bites themselves rarely cause death, but they sometimes cause a disease called rat-bite fever. The psychological trauma of the bite itself can also be significant.

Only a few types of rodenticides are used against commensal rodents. Most common are the anticoagulants (including warfarin, pindone, diphacinone, and fumarin), which are used against rats. These are used primarily by commercial pest control operators. The baits must be replenished frequently and require about a week of continuous feeding for the kill, which usually makes them impractical for use by nonprofessionals.

Anticoagulants represent about 95 percent of the rodenticide market in the United States. An estimate of anticoagulant use (Mampe 1970) shows the following breakdown for 1 year:

User	Amount of concentrate (lb)
Pest control operators	600,000
Federal agencies (including military)	4,000
Farmers	76,000
Other (including over-the-counter sales)	180,000
Total U.S. market	860,000

The concentrate marketed is generally 0.5 percent, except for diphacinone, which is marketed as 0.1 percent concentrate. According to NPCA figures, the pest control operator market accounts for 70 percent of the U.S. anticoagulant market. It is estimated that currently about 15 percent more anticoagulants are used than in 1970. The highly toxic zinc phosphide is also used occasionally by rat control specialists, generally in initial cleanouts of rodent infestations.

House mice are also important household pests, but there is no really effective chemical or nonchemical method of control available. Before the cancellation of most DDT registrations, 50 percent DDT dust was the chemical most often used against mice.

Pest control operators service industrial and commercial customers, especially food-handling firms, warehousers, and shippers, as well as residential clients. Industrial losses from rodents include consumption of food, contamination of food, and gnawing damage. No breakdown between commercial and residential rodent control is available.

Local departments of health may support rat control programs, either funded in part by the federal government or wholly locally funded. In 1969, the federal government began the Urban Rat Control Program under the Department of Health, Education, and Welfare. Appropriations have been at the level of $15 million per year. Federal money is generally used as seed money to start the program, but the local government soon takes over full responsibility. Today, 39 cities have projects that emphasize sanitation and hygiene as well as killing rats with chemicals.

In the government-run projects, in contrast to commercial pest control operations, red squill is nearly always the chemical of choice. Red squill is a relatively safe chemical that acts as an emetic and kills the rats rapidly. The Federal Working Group on Pest Management developed a list of pesticides proposed for use in 1972. (These are maximum figures; an agency might use less than the proposed amount. If it wished to use more, it would ordinarily amend its application during the year so that this extra amount would appear on the list.) The following rat control chemicals are listed (data from Midwest Research Institute 1974):

Pesticide	Pounds of Active Ingredient
Red squill	12,075
All anticoagulants	274
Warfarin	200
Diphacinone	35
Pindone	22
Fumarin	17
Zinc phosphide	2,449

Field Rodents

Field rodents are pests primarily in the western states. The rodenticides most widely used against them are strychnine and sodium fluoroacetate (1080). The Federal Working Group data cited above lists the following proposed amounts for these two chemicals:

Pesticide	Pounds of Active Ingredient
Strychnine	15,486
Sodium fluoroacetate	10

Additional data are available from EPA on the estimated acreage treated and pounds of chemicals used for field rodent control by the federal government on public lands. Since 1971, however, as a result of the controversy over the environmental effects of the chemicals, the federal government has not used any rodenticides at all on public lands in the West.

Structural Rodents

These rodents are primarily those that infest grain elevators. These, too, are of most importance in the western states. Few data are available on the amounts of chemicals used to control these pests.

THE ECONOMICS OF PEST CONTROL FOR PUBLIC HEALTH OBJECTIVES

Good economic analyses of public health programs are almost nonexistent because of two interacting factors: (a) a general failure of the public health specialists to assess directly trade-off questions (e.g., the associated costs and benefits of larger or smaller programs), and (b) a failure of economists to deal adequately with several issues such as the benefits or value of health (versus illness), the value of human life, or even the value of having fewer pest mosquitoes about. In addition, there continues to be great uncertainty about the technical feasibility of various kinds of control mechanisms. Even unit-cost coefficients for such standard vector control devices as source reduction, larviciding, and adulticiding vary greatly by region and are not readily available without a major effort in data collection.

Economic analyses can and should be brought to bear on a number of issues in the vector-control field. Hence this section identifies the magnitude of current expenditures on vector control and provides case examples of alternative economic means for achieving given control levels assessed in terms of market prices. However, with respect to the optimal amounts of resources that "should" be diverted to vector control or the social (versus market) costs of achieving specified targets, there are no studies to report or analyze.

Control Programs

Pesticide use for public health purposes accounts for only a small proportion of total pesticide expenditures. Similarly, total expenditure on control of public health vectors is fairly small relative to total public health costs. In California, for example, counties spent approximately $100 million in 1972-1973 for public health purposes, 30 percent of which went for environmental health. These data contrast with the approximately $10 million spent by California mosquito abatement districts. In Florida, too, mosquito control districts spent about $10 million, compared with county health department expenditures of more than $30 million. Internationally, one of the largest organized efforts in pest control is the malaria program of WHO. In 1971 the cost of these projects totaled less than $6 million, although the resources spent by the countries themselves were much greater, perhaps as much as $350 million. Nevertheless, in relative terms, the amounts of funding specifically designated for public-health vector control are small.

Whether the amount expended on vector control is "too large" or "too small" is a question the study team can only raise. It may be instructive, however, to indicate what would be required for an answer, if only to highlight the problems of analysis and the additional data that would be needed.

Analytical Needs

In the case of mosquito control, there is need for the establishment of threshold values at which the incidence of mosquito infestations, for example, becomes serious from a public health or nuisance point of view. For mosquito-borne diseases (e.g., yellow fever), the acceptable threshold may be very low; in the case of nuisance mosquitoes, the threshold level of control deemed "necessary" may vary greatly according to human population densities, the type of mosquito, the history of mosquito infestations in a given region, and the clientele's expectations.

Thresholds below which mosquito densities must be reduced in order to prevent transmission of vector-borne disease have been assigned by rule-of-thumb in several cases. The longevity of the residual effectiveness of a wall deposit to ensure nontransmission of malaria is assessed on the assumption that one anopheline bite per man-night is the threshold. With light-traps in California, the risk of transmission of western equine encephalitis is considered by W. C. Reeves and his research group to become nil when the catches of female *Culex tarsalis* fall below 10 per trap-night. With yellow fever in Africa, experience from the 1965 epidemic in Senegal (Chambon et al. 1967) indicates that urban transmission of the virus will occur in communities with a Breteau index of the vector *Aedes aegypti* in excess of 50 (i.e., 60 containers positive for larvae per 100 houses) and that urban epidemics cannot develop when this index is less than 5 (Pichon et al. 1969).

For nuisance mosquitoes, the aim of the activities of the mosquito abatement districts is to reduce densities to an "acceptable" level, the implication being that the contributing public gets the degree of control that it pays for. Thus, on-line assessment of the current situation is based on the number of complaints or service requests per week. It has been established in California that this "telephone index" correlates well with the assessment resulting from a canvass of individual citizens; its correlation with the actual mosquito population as assessed by light-trap counts is still under study. The mosquito abatement districts in California do have a program for standardized larval and adult density assessment, including the measurement of landing rates and biting rates (Mulhern 1973).

Tolerance of domestic animals to pest mosquitoes is an important component in assessing benefits of control, though less well understood. Quantitative studies in North America are thus far limited to several analyses involving beef animals. Steelman and his group have shown, for example, that the effect of populations of 6-115 feeding females of *Psorophora confinnis* per square foot of barn wall will cause a reduction in average annual weight gain of 18-45 pounds per animal (Steelman et al. 1972).

While physical measurement problems are difficult, the problems of economic analysis are even more severe. For example, if a 20 percent increase in mosquito landing rates is permitted, how are the increased human costs to be measured? In the case of public health vectors, what is the cost of a day's sickness, or, indeed, the cost of a human life in varying situations? In economic terms these are difficult questions, particularly where there is surplus labor, such as in India, Bangladesh, or Java. In these regions, there is clear evidence that increased labor supply will have little impact on the output of goods and services--the traditional measure of "benefits." Compared with agricultural uses of pesticides, where most benefits can in principle be captured through changes in yields, the economic aspects of pesticide use for public health purposes are inherently more complicated.

A second difficulty is related to technical questions that connect the amount of pesticide (or other control device) to the incidence of mosquitoes (or rats, or lice). Many of these relationships are not well understood, making it extremely difficult to calculate the relative advantages of varying levels of applications.

There is a further difficulty in analyzing investments in surveillance and maintenance in areas where eradication is presumed to have been completed. In Colombia, for example, *A. aegypti* was eradicated between 1952 and 1960, except in one city on the Venezuelan border (Groot 1972). As a consequence, surveillance was relaxed. However, by early 1971, surveys showed *A. aegypti* larvae in 15.1 percent of the houses in Barranquilla and in 33.4 percent of the houses in Cartagena. By the second half of 1971, Colombia experienced a dengue epidemic involving about 500,000 persons. It is clear that it was unwise to have relaxed surveillance after 1960. Given the data of 1960, however, it is less clear how the probability of a recurrence of the *A. aegypti* (and hence the costs) could have been estimated.

The foregoing difficulties arise even in attempting to use market prices in calculations, but when externalities and social issues are also included, the situation is further worsened. For example, suppose that with a greatly increased pesticide application, the number of mosquitoes may be reduced, thereby giving benefits to many residents in the form of reduced nuisance. Assume further that with the increased application, the likelihood of one or more serious sicknesses (perhaps deaths) from human intoxication by the pesticide is increased and that the pesticide may also enter the "food chain" with cumulative results such as the endangering of a species of bird.

How are the various alternatives to be weighed in social terms? That these are real issues can be seen from a recent Florida example. As a consequence of draining impoundments to recreate the habitat of the dusky seaside sparrow (an endangered species), there was an immediate resurgence in the number of nuisance mosquitoes in the area. This example points up some of the problems inherent in social cost-benefit analysis within a public health context.

TABLE 2-5 Present Value of the Costs of the Alert Phase at the High Schedule of Effort (at 10% Annual Discount Rate)

Item	United States, Puerto Rico, Virgin Islands	South America		Caribbean	Mexico and Central America		Total
		Areas Free of *A. aegypti*	Areas Currently Infested		Areas Free of *A. aegypti*	Areas Currently Infested	
Thousands of premises	14,500	24,700	3,400	4,700	3,800	1,000	52,100
Man-years per thousand existing premises	2.34	0.26	2.62	2.68	0.27	2.68	--
Allowance for materials, overhead and suprastructure (%)	15	22	18	24	20	22	--
Equivalent man-years per thousand existing premises	2.69	0.32	3.09	3.32	0.32	3.27	--
Equivalent man-years effort	39,010	7,900	10,510	15,620	1,220	3,270	77,530
Pay in U.S. dollars per man-year	4,580	540	1,110	330	760	510	--
Cost in millions of dollars	178.6	4.3	11.7	5.2	0.9	1.7	202.4

SOURCE: Little 1972:78.

Least-Cost Alternatives

There is one class of question, however, on which economic analysis is making some headway, namely, least-cost alternatives for reaching certain goals. Although less general than the question of optimal allocation of resources, the least-cost question is nevertheless significant. It is also more amenable to traditional kinds of economic analysis. With this approach, a *target level of control* is taken as given, and the cost-effectiveness procedure then attempts to ascertain which alternative control mechanism is best. Even in this more limited case, data requirements are still substantial (though many of the most difficult issues surrounding benefits can be avoided).

Eradication vs. Vaccination: Case Study

A case study, the best analysis known to the Study Team, will illustrate the problems and potential of such work. One central question in many mosquito-related programs is whether to seek eradication, or whether vaccination offers an alternative public health approach that is more economical. An Arthur D. Little study (Little 1972) addresses these questions directly for yellow fever in the Americas, a disease which is transmitted by *A. aegypti*.

The basic data for two approaches to yellow fever control are indicated in Tables 2-5 and 2-6. All costs are shown on a present value basis, i.e., future costs have been discounted at an annual rate of 10 percent. To be sure, a number of heroic assumptions were necessary in assembling the data on each alternative, not all of which the Study Team necessarily accepts. (In particular, there are a number of specific problems with the benefit calculations, e.g., no benefit allowance is made for the possible avoidance of dengue fever epidemics.) What is more important, however, is the analytical approach outlined in the report.

For example, the importance of one critical parameter, the discount rate, is illustrated in Figure 2-3. At low discount rates, yellow fever eradication clearly dominates. It is only at discount rates greater than 10 percent and only when South American vaccinations alone are deemed "adequate" that eradication is the poorer alternative. On the other hand, a high discount rate may indeed be appropriate for Latin America, making the two approaches valid alternatives.

Another interesting aspect of the Little report is the attempt to quantify benefits from the yellow fever control.

TABLE 2-6 Present Value of the Costs of Vaccination (in $ million) (at 10% Annual Discount Rate)

Type of Vaccination	United States, Puerto Rico, Virgin Islands	South America	Caribbean	Mexico and Central America	Total
Mass vaccination of initial population	34.3	45.0	6.5	15.6	101.5
Revaccination of initial population	23.3	78.9	10.6	25.7	138.5
Vaccination of newborns	23.5	216.3	28.4	82.4	350.6
Revaccination of newborns	6.3	57.0	7.6	22.2	93.1
TOTAL	87.4	397.3	53.1	145.9	683.7

SOURCE: Little 1972:99.

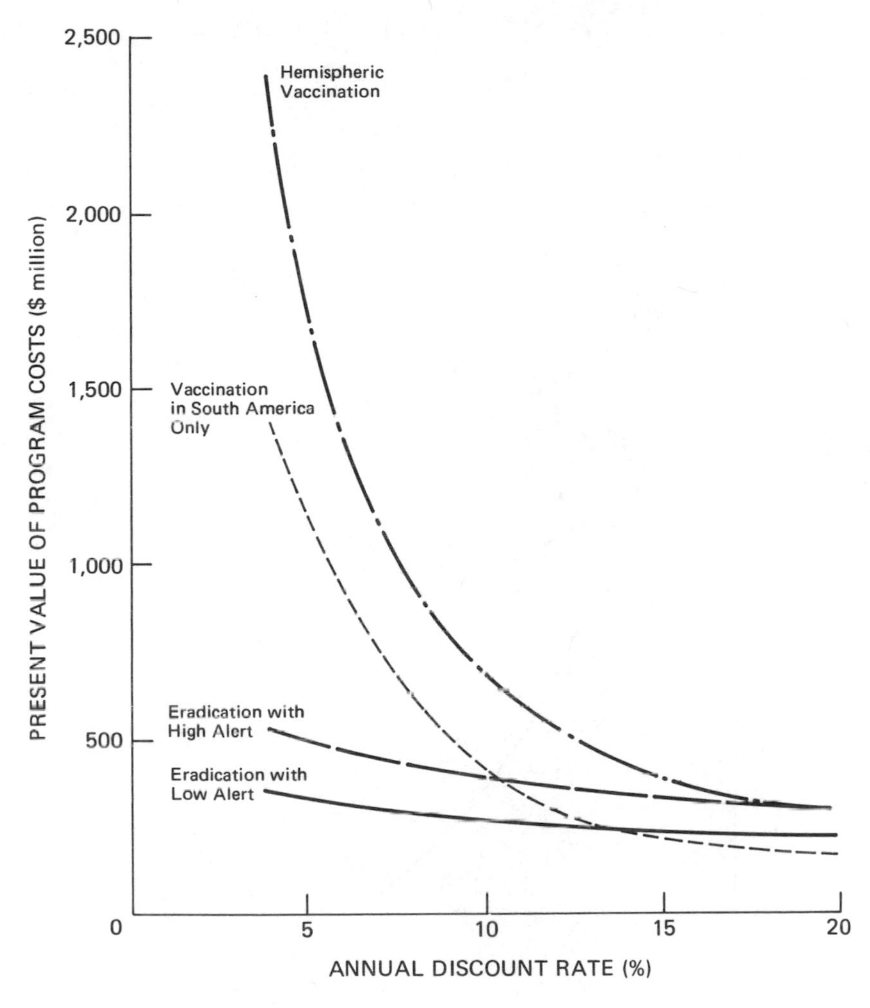

FIGURE 2-3 Comparison of eradication and vaccination programs on a cost basis (yellow fever).

Source: A. D. Little 1972:104. The Prevention of Diseases Transmitted by *Aedes aegypti* in the Americas--A Cost-Benefit Study. (Prepared for the Pan American Health Organization).

Without substantial tourist benefits, costs exceed benefits at all discount rates of greater than 6 percent (see Figure 2-4). This conclusion raises two specific questions: Are tourist decisions closely related to yellow fever outbreaks? Does either approach to yellow fever control make sense? The recent experience in Trinidad, where a small

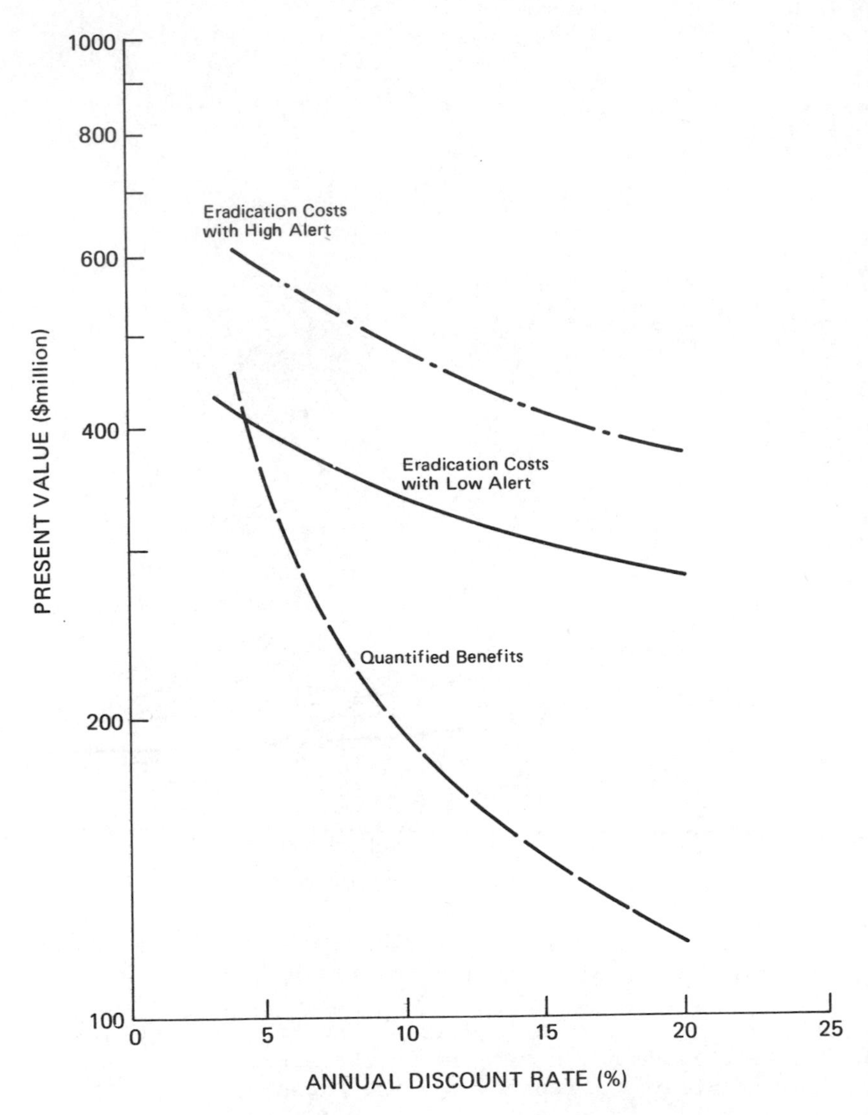

FIGURE 2-4 Comparison of costs and quantified benefits for the western hemisphere (yellow fever).

Source: Little 1972:161. The Prevention of Diseases Transmitted by *Aedes aegypti* in the Americas--A Cost-Benefit Study. (Prepared for the Pan American Health Organization).

number of cases caused great damage to the tourist trade suggests that vector control inaction may indeed be very costly to an economy over and above the direct costs of sickness. More generally, Figure 2-4 indicates the sensitivity of results to "benefits" calculations and some of the problems of estimating this side of the cost-benefit equation.

The significant point about the Little report, however, is not so much the particular results, interesting though they are. Rather it is the approach, which looks at the stream of costs necessary to accomplish a given objective. [The necessity of using streams of costs and benefits in the analysis, and the sensitivity of results to discount rates, are also the key conclusions of Cohn's antimalaria studies for India (Cohn 1973).] What is required more generally in the public health field is an assessment of the costs of achieving specified target levels for the alternative control devices. Only with that information, which is likely to be generated only by greater interaction among entomologists, public health specialists, and economists, will it be possible to reach any general conclusions about the economic efficiency of pesticide use versus other approaches to vector control.

ENVIRONMENTAL IMPACT OF PUBLIC HEALTH PEST CONTROL

Pest control activities for public health objectives are certain to have an impact on the environment. Much of the arthropod-borne disease problem occurs in man-made environments--domestic, urban, or rural--where it is pointless to debate the "environmental" impact of vector control, if for no other reason than that simple sanitation, which can so often be the remedy, is unquestionably an improvement in environment. It is impacts on natural environments that are of heightened concern today, especially aquatic environments. Terrestrial environments are seldom disturbed for the control of disease vectors, with the exception of pesticide pollution in some cases or the entrance of pesticides into terrestrial life through the food chain.

Wetlands give rise to many of the public health pests originating on wild terrain. Wetlands are complex and often fragile environments that benefit mankind well beyond the understanding or appreciation of most people. In the light of current knowledge, these marshes, swamps, backwaters, and bogs can no longer be considered wasteland nor even just wildlife habitat, the favored role that for so long has been defended against certain mosquito-control practices. Today, wetlands are protected for their total roles in ecosystems: (a) their primary productivity; (b) their special production of life forms of specific value to man; (c) their role as the habitat of endangered species of plants and animals; (d) their function as nutrient traps and thus purifiers of waters; (e) their production of nutrients for adjoining lakes, estuaries, and seas; (f) their action as flood control agents; (g) their work as absorbers of damaging storm energies; (h) their role as weather modifiers; (i) their production of recreational and esthetic values; and, probably most important of all, (j) their simple preservation of precious water. The control of biting pests can have an impact on any or all of these values.

The specific impact of vector control on the environment has been little reviewed as such (Provost 1972a,b), but there is mounting evidence on all continents that vector control managers are seriously considering environmental quality. The majority of mosquito control directors in this country are aware of the problem, try to live with restraints being imposed by other public bodies, and are anxious to seek accommodation beneficial to all, however more complex and difficult their work has become. Many are leading the field in the promotion of multipurpose wetland management and in research to develop the required technology.

This leadership is best exemplified in the long history of dialogue between mosquito control workers and managers of wildlife. In the late thirties, wildlife interests were voicing opposition to the mosquito control ditching in the Middle Atlantic states. The confrontation was resumed after World War II and exacerbated by the advent of massive larviciding with DDT and other new organic insecticides. Fortunately, the confrontation evolved into collaboration, typified by collaborative studies in New Jersey, Maryland, and Delaware in the early fifties, and continuing to this day (e.g., Ferrigno and Jobbins 1966, Ferrigno et al. 1969, Springer 1964). This collaboration reached a peak in a Symposium on the Coordination of Mosquito Control and Wildlife Management held at the U.S. Department of the Interior, Washington, D.C., on April 1-2, 1959 (Springer and Vannote 1961) and sponsored by the AMCA, the Bureau of Sport Fisheries and Wildlife of the U.S. Fish and Wildlife Service, The Wildlife Society, the USDA, and the USPHS. Out of it came the National Mosquito Control--Fish and Wildlife Management Coordination Committee, which was soon emulated by similar state and regional committees that sponsored similar symposia and conferences.

In the United States, arthropod vectors of disease that are not of wetland origin (e.g., ticks, mites, fleas, eye gnats) are either not controlled or else are controlled in a manner not affecting natural environments. Biting flies developing in waters or wetlands therefore remain the chief concern.

Source Reduction

The production of mosquitoes on natural terrain can always, in theory, be eliminated by manipulation of water. None of the other biting insects generated in problem numbers in natural environments are likely to be controlled by source reduction. When the larval stages are soil organisms, as are horseflies (Tabanidae) and biting midges (Ceratopogonidae), only drainage, flooding, or massive disturbance of the soil could reduce their production. There are good economic reasons for not doing this in freshwater areas. If it were done, larval dispersion being what it is, it would have to be over vast areas of high-water-table land, and it would damage or completely destroy a host of plants and animals whose habitat is soggy soil. In Florida some salt marshes are impounded primarily for sandfly control, and some northeastern states are planning marsh flooding for greenhead (a tabanid) control. Blackflies (Simuliidae) breed in fast-flowing streams that are not usually amenable to manipulation, even if it were desirable. Thus only mosquito control practices have appreciable environmental impact.

Mosquito source reduction by simple drainage (see the discussion of environmental sanitation) is an old practice that has played a considerable role in eliminating malaria from the United States. Both vectors, *Anopheles quadrimaculatus* in the East and *A. freeborni* in the West, are far-ranging mosquitoes that breed on wild as well as disturbed aquatic terrain. Whether the antimalaria campaigns overdrained is a moot and now irrelevant question. The role of swamps and marshes in nature's economy is better understood now than it was at the turn of the century. Wetlands can no longer be drained without strong justification and even then, only after elimination of all practical alternatives, whether the mosquitoes are disease vectors or not. Mosquito abatement agencies throughout the country must now seek the permission of many state and local agencies involved with the environment before they can dig ditches if any appreciable drainage of natural wetlands will result. In coastal areas any wetland disturbance is likely to require federal permits as well, since navigable waters may be affected.

Any manipulation of water in a wetland is certain to have an ecological impact. The trend in mosquito control is, accordingly, to minimize the impact by manipulating water as little as possible. Thus, on tidelands, "open marsh water management" (Ferrigno and Jobbins 1968) becomes the objective, and ditches are dug only where needed to connect mosquito breeding depressions to tidewater or to ponds, while permanent ponds and pools are saved and isolated from the ditching system. Where the tidelands are impounded for mosquito control, the trend is to provide sufficient water control structures to permit exchange of impoundment and estuarine waters, when this will not jeopardize mosquito control (Provost 1968, 1974), so that the natural vegetation may be saved, aquatic life many move freely between marsh and estuary, and nutrients from the marsh may enter the estuary where they are needed.

In a recent assessment of stresses imposed on the salt marsh ecosystem, Cooper (1974) concluded that mosquito control ditching on the high or irregularly flooded salt marsh may enhance estuarine productivity, while diking and impounding for mosquito control may create an ecosystem more productive than the natural one. Other ecologists view mosquito source reduction on the salt marsh with less felicity, largely because so much low marsh was unnecessarily ditched in the past. Most agree, however, that definitive research is needed on ecological impacts of altering the high marsh for mosquito control.

In noncoastal areas, attempts are now commonly made to circumvent mosquito breeding without resort to the simple expedient of draining. The Tennessee Valley Authority has an outstanding record for controlling vector and nuisance mosquitoes by water-level fluctuations and other practices in water and shoreline management on its many reservoirs (TVA and USPHS 1947, Gartrell et al. 1972). In Utah, mosquito control and wildlife interests have for decades collaborated in marsh management by shoreline modification (Rees and Winget 1968). In Texas and other parts of the Southwest, playa lakes are preserved without losing water by various bottom and vegetation manipulations (Owens et al. 1970). Probably the most useful and widespread technique for mosquito control without drainage is to connect breeding areas by ditches to one or more permanent pools in order to assure constant supply of fish and other predators. This concept was originated by an Illinois pioneer in mosquito control (Clarke 1938) and remains useful, in either fresh or salt waters, to this day.

Pesticides

Pesticide use is always liable to create two possible environmental hazards: destruction of nontarget organisms and deposition of residues that magnify in food chains and eventually injure predatory animals at the top, including man. Public health pesticides are no exception, although it is noteworthy that most biting insects, especially mosquitoes, are killed at lower insecticide dosages than the majority of agricultural pests. The assessment of any pesticide's potential for environmental damage must be made individually, because each has a specific dosage requirement against each pest species and a specific impact on each nontarget organism. Environmental hazards in public health use of pesticides can be considered to fall into three categories: (1) intradomiciliary use, (2) larviciding, and (3) adulticiding.

Intradomiciliary Use

Before the advent of DDT, spraying home interiors with various pyrethrum formulations rid homes of many biting nuisances and disease vectors. It is thought to have been a major factor in eradicating malaria in the United States and was shown greatly to reduce malaria rates abroad (Russell et al. 1943). DDT ushered in the residual spraying technique that has eradicated or reduced arthropod-borne diseases in many parts of the world. Indoor use of insecticides poses no threat to outdoor environments and is totally acceptable ecologically. It bears emphasizing that the intradomiciliary use is the major use of DDT in public health programs worldwide.

Larviciding

Organochlorine larviciding for the control of biting midge and horsefly larvae required fairly high dosages and killed many nontarget organisms, as well as leaving residues in the soil. Thus, when dieldrin was used at 1 or 2 pounds per acre against salt-marsh sandflies, the destruction of marine fish and invertebrates, other than mollusks, was nearly total (Harrington and Bidlingmayer 1958). Yet the practice was in vogue for many years in Florida and throughout the Caribbean area. It was abandoned only when it was realized that the insects became highly resistant to organochlorine compounds within a few generations. Salt-marsh tabanids were also larvicided with organochlorines, especially in the northeastern states. Results were indifferent, but only public criticism of such insecticide use forced the termination of most programs. Recent research indicates excellent control of salt-marsh sandflies, but not tabanids, with certain organophosphate insecticides at dosages demonstrated not to harm nontarget animals (Wall and Marganian 1973).

In the early years of DDT use as a larvicide against mosquitoes, considerable damage was done to nontarget organisms. This was especially true in salt marshes, where important marine crustacea, e.g., shrimp, crabs, and copepods, were affected. By the early fifties, resistance to the use of chlorinated hydrocarbon insecticides was widespread, and their use was eliminated or curtailed in the South and West but not in the northern states. Repeated applications resulted in mounting residual deposits of DDT: in some northeastern salt marshes, as high as 20 pounds per acre. In time, food-chain effects were likewise reported. In 1966 an environmentalist group sued a mosquito control commission over misuse of DDT. It was the beginning of a series of court actions and hearings that culminated in the ban on all DDT use in the U.S.

Resistance to the first class of organic compounds forced mosquito control to turn to organophosphate larvicides, and these are still in common use. Their accumulation in the environment is minimal (Ludwig et al. 1968). Although the earlier organophosphates, like parathion and malathion, can be hazardous to nontarget aquatic life, several of the more recently developed compounds, like Dursban and Abate, have relatively little environmental impact. The selective use of organophosphates is on the whole more ecologically acceptable than the use of organochlorine compounds, but mosquito resistance has become widespread. Larvicides now under development are being screened carefully for nontarget effects and some are proving relatively safe, e.g., the synthetic pyrethroids and the growth inhibitors. Paris green is the most selective larvicide for mosquitoes ever used and has a perfect record of safety in the environment; its apparent demise is unfortunate. Finally, a return to straight petroleum formulations is also under way, with considerable assurance of environmental safety (Mulla et al. 1971) and of low risk of resistance.

Blackfly (Simuliidae) control by application of DDT to streams was practiced for many years in many parts of the world in spite of proven damage to nontarget lotic life (Corbet 1958, Jamnback and Eabry 1962). The shift to the much safer methoxychlor occurred only after it had been demonstrated in 1964 that DDT accumulations in fish living in lakes receiving treated streams were interfering with their reproduction (Burdick et al. 1964). Today larviciding with Abate is the method of choice in blackfly control for reasons of both effectiveness and environmental safety. It was effective in California at dosages as low as 0.1 ppm, allowed to flow in for an hour, without harm to the rare Mojave chub, *Siphateles mohavensis* (Pelsue 1971).

Adulticiding

The practice of adulticiding and residual spraying by outdoor space treatments was made possible by DDT, which has been widely used in mosquito, sandfly, and blackfly control as a spray and as an aerosol fog, from the ground or aerially. Although charges of environmental damage have been few, and these mostly by beekeepers, there have really been very few investigations of possible damage to nontarget organisms. Dosages have always been low in this operation, whether the insecticides used were organochlorine or organophosphate compounds. Nevertheless, effects on nontarget animals are certain to occur, but if they are not spectacular, they will go unnoticed unless they are sought for. A spectacular example occurred in California in 1968-1969, when several years of fogging a city caused severe damage to surrounding pine forests by an explosive increase in scale population attributable to a derangement of the parasite-predatory complex (Garcia et al. 1972).

The present trend in adulticiding is toward the ULV method, which disperses the insecticide as a concentrate but at extremely low dosages. None of the chlorinated hydrocarbon insecticides and only four organophosphates have so far qualified for this usage. ULV treatments are apparently effective against a wider spectrum of biting insects than conventional fogs or sprays. There have been few reports on the environmental impact of ULV spraying, even though it has been widely practiced for many years. Joseph et al. (1972) found ground ULV malathion to be harmless to goldfish, mice, and quail. Pesticidal drift may be an especially acute problem with ULV in some situations.

Summary

The shift to more environmentally safe insecticiding practices in public health pest control has, historically, been precipitated by insecticide resistance or by public protest of one kind or another. Today, however, in both research and application, concern for environmental implications is a normal part of the development of pest control practice.

Exotic Larvivorous Fish

Until recently, the introduction of exotic (nonnative) larvivorous fish for the control of mosquitoes was considered altogether unobjectionable. Advances in ecological knowledge, particularly in aquatic systems, are now exposing areas of concern that demand that this form of biological control be attended by more vigilance (Bay 1973).

The introduction of exotic larvivores may pose a serious danger to the local, native fish, both small and large (Myers 1965, WHO 1973: 65-66). In natural waters all ecological niches are usually filled, so that ". . . a successful introduction is not only a faunal addition, it is also a substitution for a fraction of the native biome" (Hubbs 1972). The kind of fraction displaced and its dimension are the problems, and how this process occurs is the clue to its amelioration if the introduction is still considered desirable. Unfortunately, there is no real body of careful research on such an impact. The principal danger concerns the possible reduction in abundance, or local elimination, or total extinction, of small fish and invertebrates of scientific interest and value greater than is locally understood. Considerable international effort has lately been expended in the attempt to save rare and endangered species of animals and plants. Some of these very species are among those that may be adversely affected by the introduction of mosquito fish (*Gambusia affinis*), guppies (*Poecilia reticulata*), and other exotic larvivores (Smith 1973b).

Another danger in introductions lies in the ability of the exotic fish to alter materially the aquatic ecosystem by their severe predation while drastically modifying food chains and changing the relative abundance of many biotic elements. This often seriously affects such basic parameters as the balance between zooplankton and phytoplankton. Although not much investigated in natural waters, these effects have been demonstrated in aquaria and outdoor experimental pools, where it has been shown that *Gambusia affinis* introductions can in some situations bring about eutrophication (Hurlbert et al. 1972), a deterioration of water quality usually ascribed simply to an oversupply of nutrients.

Exotic fish brought into a climatically suitable area, even if the intent is to keep them only in artificial ponds, almost invariably escape into other local waters in a relatively short time when heavy rains cause these ponds to overflow or the pond walls are accidentally breached. Once in local streams, they eventually invade wide areas, as the unwanted carp (*Carpinus carpio*) has so well demonstrated in the United States. Moreover, unless the species cannot compete with the local fish fauna and is soon eliminated, a fish introduction cannot be undone, or repaired, like a chemical accident. An introduction is permanent, and its impact on the native fish and their environment goes on indefinitely. Under these circumstances a decision to introduce an exotic fish, even a small larvivore, can no longer be taken lightly.

The influence of this modern assessment of larvivore introduction is visible on all sides. In addition to a more demanding characterization of suitability of fish for introduction, practitioners of biological control have developed detailed protocols for the intelligent use of fish in mosquito control (Smith 1973b). Research, whether into the refinement of technology with a long-established agent like *Gambusia affinis* (Hoy et al. 1971) or into the development of a new agent like the annual fishes of the genus *Cynolebias* (Hiscox et al. 1974), places the introduction in its total environment, assessing all possible impingements on the ecosystem. This enlightened approach in research is gradually permeating the actual practice of larvivore introduction.

MOSQUITO CONTROL AND SHRIMP PRODUCTION

Commercial shrimp have been the most valuable fishery in this country since 1962 (Idyll et al. 1967). The 1972 and 1973 landings were worth $193 and $219 million, respectively, well above the second- and third-place salmon and tuna fisheries and representing approximately a quarter of the combined worth of all fish, crustacean, and mollusk fisheries in the United States (National Marine Fisheries Service 1974). In 1967 shrimp became this country's first 100-million-dollar fishery, and in 1973 it became the first billion-dollar fish industry (Roedel 1973). The shrimp fishery is concentrated in the Gulf of Mexico, which accounts for at least half of the nation's shrimp landings by weight and over three-quarters by value.

Three penaeid species dominate the fishery: *Penaeus setiferus*, the white shrimp; *P. aztecus*, the brown shrimp;

and *P. duorarum*, the pink shrimp. Other species, with quite different life histories, are fished commercially in New England and on the Pacific coast, but they account altogether for less than 10 percent of the total shrimp fishery's value. Although all three species occur inshore and offshore along the entire southeastern coast, white and brown shrimp dominate the fishery from North Carolina to northeastern Florida, the pink is the main form in southern Florida, while the white shrimp predominates in the Louisiana fishery, and the brown shrimp is most important in Texas. Shrimp caught in estuaries are juveniles of these penaeid forms or adults of other species that together are called "bait shrimp." These estuarine shrimp may be of some importance in local fishery economies, but they are not comparable to the great food shrimp or "commercial shrimp" industry.

Commercial shrimp are harvested by trawler well out at sea, but their life cycle contains an inshore period that brings them into contact with all the hazards associated with land masses dominated by man. No habitat in the United States is more beleaguered by hostile developments than the estuary, which is precisely where commercial shrimp must pass the most important growth phase of their lives. Dredge-and-fill operations for land development, pollution, channel cutting for transport of oil production machinery, and alteration of currents by causeway construction and ship channels are but a few of the threats to the welfare of juvenile shrimp. To this list must be added mosquito control on the tidelands bordering the open estuaries.

There appears to be no evidence to document any damage to commercial shrimp fisheries attributable to mosquito control activities on the tidelands of the Atlantic and Gulf coasts of the United States. The possibility of damage exists, nevertheless, but may not be documentable because of the great complexity of the impingement of mosquito control on shrimp production. This can be appraised only after a consideration of when and where shrimp are vulnerable to mosquito control side effects.

Penaeid Shrimp in the Estuaries

Commercial shrimp spawn in the Atlantic Ocean or Gulf of Mexico: the well-studied southwest Florida pink shrimp (Eldred et al. 1961) spawn in an area northwest of the Dry Tortugas. In 2 weeks, the eggs hatch into the first of 11 larval stages that in the ensuing 3 weeks grow to the first of several postlarval stages. While in these early stages, the shrimp move eastward approximately 100 miles toward and into the mangrove estuaries of Florida's southwest coast. Even though mortality during this migration is extremely high, the populations of young shrimp entering the estuaries are still very dense. As they enter these estuaries, they change from plankters to bottom dwellers. From their entrance on flooding tides until their departure on ebbing tides 2 to 9 months later, the shrimp grow from postlarvae less than ½ inch in length to juveniles 1 to 6 inches long. Once returned to the sea, they soon attain maturity and are candidates for the fishery.

It is obvious that growth is phenomenal during this period and is the main function of the period in the estuarine nursery grounds. Because this is the time and place where shrimp are likely to be affected by mosquito control, it is well to characterize these nursery areas. Williams (1955:141-142) gives an excellent description of nurseries in North Carolina:

> The areas most attractive to juvenile shrimps are in shallow estuaries which have a predominantly soft bottom. Such areas may be near the ocean, or in certain cases, they are miles from the nearest inlet. The population is usually more dense in localities near the sea than in those far from the inlets, apparently because of the relative accessibilities of the area. . . . Within the nursery areas, a number of bottom types may exist, and repeated collections show that a soft substratum of mud or fibrous peat is preferred by the young, as others have observed, but in special cases, sand or clay may provide a satisfactory substratum. Depth of water over such areas has no observed effect on the density of the population, although by the very nature of the preferred areas, the water is usually quite shallow. Within preferred areas, though, other factors modify population density. Observations show that availability of cover is one of the most essential requirements for a satisfactory nursery area. The types of cover available vary somewhat with the substratum and the salinity as well as with the geographic position. In brackish creeks which run through the marshes and have their headwaters in either pocosins or well-drained forest areas, there is a great quantity of forest litter scattered over the bottom. This material, in various stages of

> decomposition, affords one type of favored habitat. Another type of cover is provided by living vegetation. . . . Availability of food in the nursery areas is of paramount importance to the population, and the preferred types of bottom and cover, while easily appraised as important, are undoubtedly associated with an abundant food supply. The shrimp are bottom feeders. There is a general agreement among investigators that juvenile and adult shrimp eat any available organic material.

Shrimp nurseries elsewhere in the Southeast differ only in details from this description for North Carolina. The only notable divergence is along peninsular Florida, where grass salt marshes give way to mangrove tidelands and where rivers entering estuaries become few and relatively small. What pertains everywhere is that juvenile shrimp and salt-marsh mosquito larvae occupy different, although frequently adjoining, habitats. The shrimp are in permanent water, while the mosquito larvae are on the vegetated tidelands that are subject to various rhythms of submergence by tidewater.

It is especially germane to the problem that juvenile shrimp are an important trophic element of detritus ecosystems (Odum and de la Cruz 1963). Whether the tidelands are of the grass marsh type (Odum 1961) or mangroves (Odum 1970), the most important nutrients for the estuary are the foliage and other plant debris that are washed into the open waters by rain or tide from the vegetated tidelands. The juvenile shrimp on the nursery areas feed on this detritus and the organic matter, living or dead, associated with it. They are in turn preyed upon by a vast array of fish and other animals that themselves constitute a large source of food and recreation for man, both in the estuary and out at sea. Salt-marsh mosquitoes breed on the vegetated portion of the tidelands and are consequently integral elements in the estuarine ecosystem; thus their control impinges on the welfare of shrimp.

Potential Hazards in Salt-Marsh Mosquito Control

Each of the four methods of controlling salt-marsh mosquitoes can have an impact on shrimp production.

Dredging and Filling

This method of controlling salt-marsh mosquitoes is the most effective simply because it totally eliminates the breeding habitat. However, it deals shrimp a devastating double blow. The dredging destroys the nursery by removing the rich organic substrate, with whatever submerged cover plants it supported, leaving behind a much deeper and totally inorganic bottom. The excavated soil is then placed on the marsh, abolishing it as a marsh and destroying the plant community that contributed the vital nutrients to the estuary and its shrimp nurseries. Such an operation obviously reduces the commercial shrimp fishery by whatever proportion of juveniles originated in the destroyed nursery.

Ditching

Ditching or canalling a salt marsh eliminates mosquito production (a) by assuring that no water stands on the marsh long enough to accommodate a developing brood of mosquitoes and (b) by providing minnow access to all parts of the marsh. Done properly, it does not drain the marsh but contributes to it a vast system of channels to supplement whatever natural tide channels already exist. It could harm shrimp if permanent ponds and pools once serving as nurseries are rendered temporary or are eliminated by connection with the ditch system. It could benefit shrimp if large expanses of solid marsh become interspersed with canals large enough to function as nurseries. It may also benefit estuarine shrimp nurseries by expediting the flushing of detritus off the marsh (Cooper 1974, Provost 1974).

Weirs placed in tide channels or canals to stabilize water levels are employed in fishery, waterfowl, and muskrat management on tidelands. Because the crest of the weir is usually set only a few inches below average marsh ground elevation, they serve mosquito control as well by creating semi-impoundments comprising natural ponds and pools, all connected to a permanent deep-water channel system harboring an abundant reservoir of larvivorous minnows. It has been shown that these semi-impoundments create favorable environments for juvenile shrimp (Herke 1971), so there is no question of hazard whatever.

Impoundments

An inexpensive and very effective way to stop the production of salt-marsh mosquitoes is to throw a dike around the marsh and keep it flooded, thus denying the female mosquito access to the moist soil it uses for egg deposition. If no water exchange between impoundment and estuary is provided for, shrimp may be damaged (a) by isolating whatever nursery areas--channels and pools--may have existed within the dikes before impoundment, (b) by preventing the creation of detritus if too much of the emergent vegetation is killed by submergence, and (c) by not allowing the accumulated detritus and other nutrients to reach the estuarine nursery areas. It may be possible to avoid all these ill effects by using whatever structures and schedules are necessary to open and close impoundments for the maximum benefit of estuarine conservation while also forestalling the development of mosquitoes (Provost 1968, 1974).

Larviciding

If the toxicants used against salt-marsh mosquito larvae reached only the salt marsh where the larvae are and never beyond, there would be no hazard to juvenile shrimp, because the two do not inhabit the same areas. However, aerial application of insecticides can drift, and the shallow waters around the salt marsh can be important shrimp nurseries. Even if drift does not occur, the deposited insecticides can still be washed into the estuaries with tides and rain (Butler 1968). The danger, therefore, exists that larvicides applied to a salt marsh may reach juvenile shrimp in the estuary. They do reach shrimp in channels and pools within the marsh.

It is not surprising that shrimp, being crustaceans, are sensitive to many insecticides. It was early shown that they were highly susceptible to the chlorinated hydrocarbon insecticides (Chin and Allen 1957, De Sylva 1954, Harrington and Bidlingmayer 1958, Springer and Webster 1951). Ongoing studies in Texas are demonstrating that shrimp are nearly all killed by malathion at dosages commonly used in mosquito control (unpublished data, National Marine Fisheries Service). It can be reasonably assumed that most organophosphate insecticides used in mosquito control would kill juvenile shrimp as readily as malathion.

Butler (1968:123) has well summarized the problem of insecticides and juvenile shrimp:

Juvenile animals, in general, are more susceptible than adults to pesticides, and peak populations of juveniles--both those hatched in the estuary and migrants from offshore waters--are present when pesticide pollution is usually highest, in the spring and early summer. This situation is of special concern to the commercial fishermen, since both shrimp and menhaden, respectively our most valuable and largest marine crops, migrate through the estuaries as juveniles. . . . Residues in shrimp and other crustaceans are usually negligible or absent. I suspect that this is because they are more sensitive to insecticides than other estuarine fauna and are killed when exposed to this type of pollution, before they have opportunity to metabolize the poison.

Actual Damage to Shrimp Fishery

Damage to the shrimp fishery must be considered separately from isolated instances of local damage to juvenile shrimp and their estuarine habitat. The potential damage to juvenile shrimp from mosquito-control operations may be summarized in the order of the hazards discussed above:

Dredging and filling. Florida has filled some 3,000 acres of salt marsh by hydraulic dredging for mosquito control. The practice was halted in 1967. This method was used in the Indian River region, where it probably damaged some nursery grounds of brown and white shrimp. No other state has used this method of salt-marsh mosquito control. The dredging and filling for mosquito control accounts for only a fraction of the marsh destruction on Florida's east coast, which has been done to create waterfront residential and industrial development, for which no data have been compiled; on the Gulf coast of Florida, where dredging and filling has been only for navigational, industrial and residential developments, it has been estimated that 18,406 acres of tideland (2.0 percent of total) have been destroyed (McNulty et al. 1972).

Ditching. From Texas to North Carolina, about a million acres of salt marsh have been ditched for mosquito control. Without specific investigation, it is impossible to say how much of this ditching benefited shrimp production and how much, if any, damaged it. Some ecologists estimate more good than harm (Cooper 1974). Because the most important nursery areas for the Gulf fishery are surrounded by millions of acres of still unditched tidelands in Texas, Louisiana, and Florida, existing mosquito control ditching, even if assumed to be deleterious, could not have hurt the fishery in an appreciable way. Semi-impoundments are almost totally a Louisiana phenomenon, where Herke (1968) estimates 250,000 acres have been so treated. Louisiana and other states are studying the technique for mosquito control. It is, as stated earlier, all to the benefit of shrimp production.

Impoundments. Impoundments for the control of salt-marsh mosquitoes are almost entirely restricted to Florida. They are resorted to wherever tides make ditching ineffective. Some 50,000 acres have been impounded--mostly in the Indian River region. Damage to nursery grounds of brown and white shrimp has undoubtedly resulted. Studies in tidelands management are under way to learn how to mitigate such effects (Provost 1974). Of the 6 million acres of salt marsh and mangrove swamp in the Gulf coastal states, only a few hundred are or ever will be impounded for mosquito control, so the practice is altogether inconsequential to shrimp production.

Larviciding. It is fortunate that in the heyday of DDT larviciding, mosquito control was not practiced anywhere near the most important shrimp nursery grounds. Organized mosquito control did not exist then in Texas, Louisiana, the southwest coast of Florida, northeast Florida, Georgia, and the Carolinas. Today, all these areas have organized mosquito control districts, but none applies organic insecticides to its tidelands as a common practice. The three most important shrimp fishery states--Texas, Louisiana, and Florida--have, for 10 years or more, actively and successfully urged districts not to larvicide with organic compounds. A notable exception is Lee County, Florida, which insists on a large larviciding program with Abate. Its estuaries are, fortunately, only a small component of the total nursery areas for the Dry Tortugas pink shrimp fishery.

Because the Tortugas fishery is one of the largest in the United States, information on the question of possible damage from mosquito control operations was sought from an authority on the shrimp fishery, Mr. Edwin A. Joyce, Jr., Chief, Bureau of Marine Science and Technology, Florida Department of Natural Resources. He stated that it would be extremely difficult to prove whether such damage existed because of the enormous *natural* fluctuation from year to year in all shrimp populations. The Tortugas production varies from 11 to 25 million pounds per year (average, 18 or 19). Boats are so well equipped to harvest and to communicate from boat to boat that the annual production is invariably well harvested. Marking and tracing studies have shown nursery areas for the Tortugas fishery to extend from the Florida Keys and Florida Bay through the estuaries of Everglades National Park and the Ten Thousand Islands all the way up to Sanibel Island, in Lee County. Mosquito control is practiced on but a small fraction of this estuarine system. There are so many natural factors affecting the survival of juvenile shrimp in the estuaries--predation, cannibalism, intolerable sudden changes in water temperature and salinity, and so on--that it would be virtually impossible to trace the effect of damage to one nursery area to the harvested crop off the Tortugas.

Summary

It is clear that mosquito control operations on the tidelands of the Southeast could damage the shrimp fishery. It may have done so in the past, but in a localized and minor way only. Looking at the entire fishery of approximately 200,000,000 pounds per year, it is most unlikely that mosquito control could have diminished production by more than a small fraction of 1 percent, a quantity that would be literally lost among the multitude of natural factors resulting in the great annual fluctuation in production. Authorities on the shrimp fishery are far more concerned about other activities of man on the tidelands, which pose infinitely greater threats to shrimp than does mosquito control as currently practiced. Among these activities are dredge-and-fill operations for land development and oil exploitation; channelization for navigation and for transport of oil production machinery; alteration of currents and streams by channel and causeway construction, which affects water salinity and temperature regimes; and pollution of all kinds (Futrell 1974, Kutkuhn 1966, Louisiana Advisory Commission on Coastal and Marine Resources 1973, Louisiana Wildlife and Fisheries Commission 1971, McNulty et al. 1972, U.S. Fish and Wildlife Service 1970).

REFERENCES

Ackerman, A. B. (1968) Crabs--the resurgence of Phthirus pubis. N. Engl. J. Med. 278:950-951.

American Mosquito Control Association (1972) Mosquito control in the United States: Preliminary data on area served and budgets. Mosq. News 32(2).

Anderson, D. W., and J. J. Hickey (1972) Eggshell changes in certain North American birds. Proc. 15th Int. Ornith. Cong. pp. 514-540.

Anonymous [J. A. Mulrennan and M. W. Provost] (1951) Policy of the Florida State Board of Health pertaining to present mosquito control practices and recommendations for a long-range plan to bring about more effective control in the state. Mimeo, 12 pp. Adopted by State Board of Health session, Nov. 18, 1951.

Anonymous (1966) Comprehensive analysis of the professional mosquito control market. Pest Control 34(3):69, 76.

Barber, M. A., and T. B. Hayne (1921) Arsenic as a larvicide for anophelines. U.S. Publ. Health Rep. 36:3027-3034.

Bay, E. C. (1973) Exotic fish introductions for mosquito control: possible and purported consequences. WHO mimeo VBC/WP/73.7.

Beaver, P. C., and T. C. Orihel (1965) Human infection with filariae of animals in the United States. Am. J. Trop. Med. 14:1010-1029.

Benenson, A. S. (ed.) (1970) The Control of Communicable Diseases in Man. 11th ed. New York: American Public Health Association.

Bishopp, F. C. (1940) Cooperative investigations of the relation between mosquito control and wildlife conservation. Science 92(2383):201-202.

Bishopp, F. C. (1949) Larvicides, pp. 1339-1359. *In* Malariology, M. F. Boyd (ed.). Philadelphia: W. B. Saunders Co.

Brown, A. W. A. (1972) The ecological implications of insecticide usage in malaria programs. Am. J. Trop. Med. Hyg. 21:829-834.

Bruce-Chwatt, L. J. (1971) Insecticides and the control of vector-borne diseases. Bull. WHO. 44:419-424.

Bruce-Chwatt, L. J. (1973) Global problems of imported diseases. Adv. Parasitol. 11:75-114.

Burdick, G. E., E. J. Harris, H. J. Dean, T. M. Walker, J. Skea, and D. Colby (1964) The accumulation of DDT in lake trout and the effect on reproduction. Trans. Am. Fish. Soc. 93(2):127-136.

Butler, P. A. (1968) Pesticides in the estuary, pp. 120-124. *In* Proc. Marsh and Estuary Management Symposium, L.S.U., Baton Rouge, July 19-20, 1967.

California Mosquito Control Association (1973) Year Book. (Also previous years.)

Center for Disease Control (1971) Morbidity and Mortality 20(5).
Center for Disease Control (1972) Manual for Procedures for National Morbidity Reporting. DHEW Publ. No. (HSM) 72-8113.
Center for Disease Control (1973a) Morbidity and Mortality 22(35):299.
Center for Disease Control (1973b) Morbidity and Mortality. Annual Summary, 1972.
Chadwick, P. R. (1970) The activity of de-allethrolone d-trans chrysanthemate and other pyrethroids in mosquito coils. Mosq. News 30:162-170.
Chambon, L., I. Wone, and P. Brès (1967) Une épidemie de fièvre jaune au Sénégal en 1965. Bull. WHO. 36:113-150.
Chin, E., and D. M. Allen (1957) Toxicity of an insecticide to two species of shrimp, *Penaeus aztecus* and *P. setiferus*. Tex. J. Sci. 9(3):270-278.
Clarke, J. L. (1938) Mosquito control as related to marsh conservation. Proc. N.J. Mosq. Exterm. Assoc. 25:139-146.
Cohn, E. J. (1973) Assessing the costs and benefits of anti-malaria programs: The India experience. Am. J. Publ. Health 65:1086-1096.
Cooper, A. W. (1974) Salt marshes, pp. 55-98. *In* Coastal Ecological Systems of the United States. Vol. II. Washington, D.C.: The Conservation Foundation.
Corbet, P. S. (1958) Some effects of DDT on the fauna of the Victoria Nile. Rev. Zool. Bot. Afr. 57:73-95.
Culbertson, J. J. (1944) Possible hazards from filariasis in the United States. JAMA 126:267.
De Benedetti, A. (1935) Disappearance of arsenic from the surface of water delarvaed with Paris green. Riv. Malar. 15(Sec. 1):438-447 (In Italian).
De Sylva, D. P. (1954) The live bait shrimp fishery on the northeast coast of Florida. Fla. Board of Conservation, Tech. Ser. No. 11:1-35.
Eldred, B., R. M. Ingle, K. D. Woodburn, R. F. Hutton, and H. Jones (1961) Biological observations on the commercial shrimp, *Penaeus duorarum* Burkenroad, in Florida waters. Fla. Board Conservation, Marine Lab., Prof. Papers Ser. No. 3, 139 pp.
Ellis, M. M., B. A. Westfall, and M. D. Ellis (1941) Arsenic in fresh-water fish. Ind. Eng. Chem. 33:1331-1332.
Entomological Society of Canada (1970) Pesticides and the environment. Bull. Entomol. Soc. Can. 3(1):1-16.
Ferrigno, F., and D. M. Jobbins (1966) A summary of nine years of applied mosquito-wildlife research on

Cumberland County, N.J. salt marshes. Proc. N.J. Mosq. Exterm. Assoc. 53:97-112.

Ferrigno, F., and D. M. Jobbins (1968) Open marsh water management. Proc. N.J. Mosq. Exterm. Assoc. 55:104-115.

Ferrigno, F., L. G. MacNamara, and D. M. Jobbins (1969) Ecological approach for improved management of coastal wetlands. Proc. N.J. Mosq. Exterm. Assoc. 56:188-202.

Florida Department of Health and Rehabilitative Services (1973) Annual report, Division of Health. Jacksonville. pp. 86-119.

Florida Division of Health (1973) Annual Report, 1972. Bureau of Entomology section. Jacksonville. pp. 84-116.

Fontaine, R. E., and P. E. Rosen (1973) Evaluation of Abate insecticide formulations as larvicides against *Anopheles gambiae* in northern Nigeria. Mosq. News 33: 428-440.

Freeborn, S. B. (1949) The relationship of mosquito abatement agencies to other agencies of government. Proc. Calif. Mosq. Control Assoc. 17:65-68.

Futrell, W. (1974) Oil and trouble in the Louisiana wetlands. Sierra Club Bull. 59(7):14-16.

Garcia, R., K. H. Hansgen, and F. C. Roberts (1972) Observations on malathion thermal fogging in a mixed conifer forest at Lake Tahoe. Proc. Calif. Mosq. Control Assoc. 40:56-59.

Gartrell, F. E., W. W. Barnes, and G. S. Christopher (1972) Environmental impact and mosquito control water resource management projects. Mosq. News 32(3):337-343.

Goodman, M. L., and I. Gore (1964) Pulmonary infarct secondary to dirofilaria larvae. Arch. Intern. Med. 113:702-705.

Gratz, N. (1973) The current status of louse infestations throughout the world. pp. 23-31. *In* PAHO Sci. Publ. 263.

Griffitts, T. H. D. (1927) Moist sand method of applying Paris green for destruction of subsurface feeding mosquito larvae. U.S. Public Health Rep. 42:2701-2705.

Groot, H. (1972) Report on dengue (Ministry of Public Health of Colombia). Dengue Newsl. Am. 1:3-8.

Harrington, R. W., Jr., and W. L. Bidlingmayer (1958) Effects of dieldrin on fishes and invertebrates of a salt marsh. J. Wildl. Man. 22(1):76-82.

Hayashi, A., and Y. Takama (1972) Effet larvicidal de plusieurs produits pyrethroides sur larve de mouche et de moustique. Botyu-Kagaku 37:1-3.

Hayes, G. R., Jr., and A. B. Ritter (1966) The elimination of *Aedes aegypti* infestations in Louisiana. Mosq. News 26(3):381-383.

Heal, R. (1966) Pesticide usage in structural pest control. Paper presented before Association of American Pesticide Control Officials, August 10, 1966. NPCA.

Herke, W. H. (1968) Weirs, potholes and fishery management, pp. 193-211. *In* Proc. Marsh and Estuary Management Symposium, L.S.U., Baton Rouge, July 19-20, 1967.

Herke, W. H. (1971) Use of natural, and semi-impounded, Louisiana tidal marshes as nurseries for fishes and crustaceans. Ph.D. dissertation, Dept. Zoology and Physiology, L.S.U.

Herms, W. B., and H. F. Gray (1940) Mosquito Control. New York: The Commonwealth Fund.

Herms, W. B. (1949) Looking back half a century for guidance in planning and conducting mosquito control operations. Proc. Calif. Mosq. Control Assoc. 17:89-92.

Hiscox, K. J., E. J. Kingsford, and W. E. Hazeltine (1974) Annual fish of the genus *Cynolebias* for mosquito control. Proc. Calif. Mosq. Control Assoc.

Hobbs, J. H., and C. W. Miller (1973) A field trial of Landrin as a residual house spray in El Salvador. Mosq. News 33:521-525.

Holan, G. (1971) Rational design of insecticides. Bull. WHO 44:355-362.

Hoy, J. B., A. G. O'Berg, and E. E. Kauffman (1971) The mosquito fish as a biological control agent against *Culex tarsalis* and *Anopheles freeborni* in Sacramento Valley rice fields. Mosq. News 31(2):146-152.

Hoy, J. B., and D. E. Reed (1971) The efficacy of mosquito fish for control of *Culex tarsalis* in California rice fields. Mosq. News 31(4):567-572.

Hubbs, C. (1972) The impact of fish introductions on aquatic communities. Trop. Fish Hobby, April 1972, pp. 22-28.

Hudson, B. W., M. I. Goldenberg, and J. D. McCluskie (1971) Serological and bacteriological investigations of an outbreak of plague in an urban tree squirrel population. Am. J. Trop. Med. Hyg. 20:255-263.

Hurlbert, S. H., J. Zedler, and D. Fairbanks (1972) Ecosystem alteration by mosquitofish (*Gambusia affinis*) predation. Science 175:639-641.

Idyll, C. P., D. C. Tabb, and B. Yokel (1967) The value of estuaries to shrimp, pp. 83-90. *In* Proc. Marsh and Estuary Management Symposium. L.S.U., Baton Rouge, July 19-20, 1967.

Jamnback, H. A. (1971) Blackflies of the Americas. WHO/ VBC 71.283. Geneva: World Health Organization. Switzerland. 32 pp.

Jamnback, H. A. (1973) Recent developments in the control of blackflies. Annu. Rev. Entomol. 18:281-304.

Jamnback, H. A., T. Duflo, and D. Marr (1970) Aerial application of larvicides for control of simulium damnosum in Ghana: a preliminary trial. Bull. WHO 42(5): 826-828.

Jamnback, H. A., and H. S. Eabry (1962) Effects of DDT, as used in blackfly control, on stream arthropods. J. Econ. Entomol. 55(5):636-639.

Joseph, S. R., J. Mallack, and L. George (1972) Field applications of ultra low volume malathion to three animal species. Mosq. News 32(4):504-506.

Kapoor, I. P., R. L. Metcalf, A. S. Hirwe, J. R. Coats, and M. S. Khalsa (1973) Structure-activity correlations of biodegradability of DDT analogs. J. Agr. Food Chem. 21:310-315.

King, W. V., and G. H. Bradley (1926) Airplane dusting in the control of malaria mosquitoes. U.S. Dept. Agr. Dept. Circ. 367.

King, W. V., and T. E. McNeel (1938) Experiments with Paris green and calcium arsenite as larvicides for culicine mosquitoes. J. Econ. Entomol. 31(1):85-86.

Kutkuhn, J. H. (1966) The role of estuaries in the development and perpetuation of commercial shrimp resources. Am. Fish. Soc. Spec. Publ. No. 3:13-36.

Little, A. D., Inc. (1972) The Prevention of Diseases Transmitted by *Aedes aegypti* in the Americas--A Cost-Benefit Study (Prepared for the Pan American Health Organization).

Louisiana Advisory Commission on Coastal and Marine Resources (1973) Louisiana Wetlands Prospectus. Baton Rouge.

Louisiana Wildlife and Fisheries Commission (1971) Cooperative Gulf of Mexico Estuarine Inventory and Study, Louisiana: Phase I, Area Description and Phase IV, Biology. New Orleans.

Ludwig, P. D., H. J. Dishburger, J. C. McNeill IV, W. O. Miller, and J. R. Rice (1968) Biological effects and persistence of Dursban insecticide in a salt marsh habitat. J. Econ. Entomol. 61(3):626-633.

Maegraith, B. G. (1963) Unde venis? Lancet 1:401.

Mampe, C. D. (1970) Rodenticide Use by Pest Control Operators. Presented at the Symposium on Rodenticides. New York: National Pest Control Association.

McNulty, J. K., W. N. Lindall, and J. E. Sykes (1972) Cooperative Gulf of Mexico Estuarine Inventory and Study, Florida: Phase I, Area Description. National Oceanic and Atmospheric Administration, Tech. Rep. NMFS CIRC-368.

Metcalf, R. L. (1972) Development of selective and biodegradable insecticides, pp. 137-174. *In* Pest Control Strategies for the Future. Washington, D.C.: National Academy of Sciences.

Metcalf, R. L., I. P. Kapoor, and A. S. Hirwe (1971a) Biodegradable analogues of DDT. Bull. WHO 44:363-374.

Metcalf, R. L., G. K. Sangha, and I. P. Kapoor (1971b) A model ecosystem for the evaluation of pesticide biodegradability and ecological magnification. Environ. Sci. Technol. 5:709-714.

Micks, D. W., G. V. Chambers, J. Jennings, and A. Rehmet (1967) Mosquito control agents derived from petroleum hydrocarbons. I. Laboratory effectiveness. J. Econ. Entomol. 60:426-429.

Micks, D. W., G. V. Chambers, J. Jennings, and K. Barnes (1968) Laboratory evaluation of a new petroleum derivative, FLIT MLO. J. Econ. Entomol. 61:647-650.

Midwest Research Institute (1974) Production, Distribution, Use and Environmental Impact Potential of Selected Pesticides. Draft Final Report for Council on Environmental Quality.

Moore, N. W. (ed.) (1966) Pesticides in the Environment and their Effect on Wildlife. (A supplement to the J. Appl. Ecol., Vol 3.) Oxford: Blackwell.

Mrak, E. M. (ed.) (1969) Report of Secretary's Commission on Pesticides and Their Relationship to Environmental Health. Pts. I and II. U.S. Dept. Health, Education, and Welfare. Washington, D.C.: U.S. Government Printing Office.

Mulhern, T. (1973) A Training Manual for Personnel of Official Mosquito Control Agencies. Visalia, Calif.: California Mosquito Control Association Press.

Mulhern, T. (ed.) (1973) A Training Manual for Personnel of Official Mosquito Control Agencies. Sacramento: California State Department of Health.

Mulla, M. S., J. R. Arias, and H. A. Darwazeh (1971) Petroleum oil formulations against mosquitoes and their effects on some nontarget insects. Proc. Calif. Mosq. Control Assoc. 39:131-136.

Myers, S. F. (1965) Gambusia, the fish destroyer. Trop. Fish Hobby., January 1965, pp. 31-32, 53-54.

National Marine Fisheries Service (1974) Fisheries of the United States, 1973. Current Fishery Statistics No. 6400.

Odum, E. P. (1961) The role of tidal marshes in estuarine production. N.Y. St. Conserv., (June-July), pp. 12-15, 35.
Odum, E. P., and A. A. de la Cruz (1963) Detritus as a major component of ecosystems. AIBS Bull. 13(3):39-40.
Odum, W. E. (1970) Pathways of energy flow in a south Florida estuary. Ph.D. dissertation, Inst. Marine Sci., Univ. of Miami.
Owens, J. C., C. R. Ward, E. W. Huddleston, and D. Ashdown (1970) Non-chemical methods of mosquito control for playa lakes in West Texas. Mosq. News 30(4):571-579.
Pan American Health Organization (1973) The Control of Lice and Louse-borne Diseases. Scientific Publication No. 263. Washington, D.C.
Pelsue, F. W. (1971) Black flies in the Southeast Mosquito Abatement District. Proc. Calif. Mosq. Control Assoc. 39:50.
Pichon, G., J. Hamon, and J. Mouchet (1969) Groupes ethniques et foyers potentiels de fièvre jaune dans les états francophones d'Afrique occidentals. Cah. ORSTROM, Ser. Entomol. Med. Parasitol. 7:39-50.
Provost, M. W. (1948) Mosquito control in the State of Florida. Fla. Health Notes 40(5):91-109.
Provost, M. W. (1968) Managing impounded salt marsh for mosquito control and estuarine resource conservation, pp. 163-171. *In* Proc. Marsh and Estuary Management Symposium. L.S.U., Baton Rouge, July 19-20, 1967.
Provost, M. W. (1972a) Environmental hazards in the control of disease vectors. Environ. Entomol. 1(3):333-339.
Provost, M. W. (1972b) Environmental quality and the control of biting flies. Canadian Defence Research Council. Proc. Symp. May 16-18, 1972. pp. 1-7. Ottawa.
Provost, M. W. (1974) Salt marsh management in Florida, pp. 5-17. *In* Proc. Tall Timbers Conf. on Ecol. Animal Control by Habitat Management 1973. No. 5.
Rathburn, C. B. (1966) The arsenic content in soil following repeated applications of granular Paris green. Mosq. News 26(4):537-539.
Rathburn, C. B., and A. H. Boike, Jr. (1973) Laboratory selection of *Culex nigripalpus* Theob. for resistance to Paris green. Mosq. News. 33(4):512-516.
Rees, D. M., and R. N. Winget (1968) Water management units and shore line modifications used in mosquito control investigations. Mosq. News 28(3):305-311.
Reeves, W. C. (1972) Can the war against infectious diseases be lost? Am. J. Trop. Med. Hyg. 21(3):251-259.
Roedel, P. M. (1973) Shrimp '73--a billion dollar business. Mar. Fish. Rev. 35:1-2.

Rogers, A. J., and C. B. Rathburn (1958) Tests with a new granular Paris green formulation against *Aedes, Anopheles* and *Psorophora* larvae. Mosq. News 18(2):89-93.

Rogers, A. J., and C. B. Rathburn (1960a) Improved methods of formulating granular Paris green larvicide. Mosq. News 20(1):11-14.

Rogers, A. J., and C. B. Rathburn (1960b) Airplane application of granular Paris green mosquito larvicide. Mosq. News 20(2):105-110.

Rudd, R. I. (1964) Pesticides and the living landscape. Madison: University of Wisconsin Press.

Russell, P. F., F. W. Knipe, and N. R. Sitapathy (1943) Malaria control by spray-killing adult mosquitoes: Fourth season's results. J. Malar. Inst. India 5:59-76.

Russell, P. F. (1955) Man's Mastery of Malaria. London: Oxford University Press.

Schoof, H. F., and R. T. Taylor (1972) Recent advances in insecticides for malaria programs. Am. J. Trop. Med. Hyg. 21:807-812.

Sharma, M. I. D. (1972) Residual insecticides for the control of arthropod-borne diseases: The scope and limitation of their use with particular reference to India. J. Commun. Dis. 4:208-218.

Smith, C. N. (1973a) Pyrethrum for control of insects affecting man and animals, pp. 225-241. *In* Pyrethrum, the Natural Insecticide, (J. E. Casida) (ed.). New York: Academic Press.

Smith, R. F. (1973b) Considerations on the safety of certain biological agents for arthropod control. Bull. WHO 48(6):685-698.

Soper, F. L. (1966) Paris green in the eradication of *Anopheles gambiae:* Brazil, 1940; Egypt 1945. Mosq. News 26(4):470-476.

Springer, P. F. (1964) Wildlife management concepts compatible with mosquito suppression. Mosq. News 24(1):50-55.

Springer, P. F., and R. L. Vannote (1961) Activities of the National Mosquito Control--Fish and Wildlife Management Coordination Committee. Mosq. News 21(2):158-160.

Springer, P. F., and J. R. Webster (1951) Biological effects of DDT applications on tidal salt marshes. Mosq. News 11(2):67-74.

Steelman, C. D., T. W. White, and P. E. Schilling (1972) Effects of mosquitoes on the average daily gain of feedlot steers in southern Louisiana. J. Econ. Entomol. 65:462-466.

Tennessee Valley Authority and U.S. Public Health Service (1947) Malaria control on impounded water. Washington, D.C.: U.S. Government Printing Office.

U.S. Department of Agriculture (1972) The Pesticide Review 1971. Agricultural Stabilization and Conservation Service.

U.S. Fish and Wildlife Service (1970) National estuary study. Vol. 2. Washington, D.C.: U.S. Government Printing Office.

United States Public Health Service, Center for Disease Control (1973) Public health pesticides. Pest Control 41(4):18-22, 24-26, 28-50.

Wall, W. J., Jr., and V. M. Marganian (1973) Control of salt marsh *Culicoides* and *Tabanus* larvae in small plots with granular organophosphorus pesticides, and the direct effect on other fauna. Mosq. News 33(1):88-93.

Williams, A. B. (1955) A contribution to the life histories of commercial shrimps (Penaeidad) in North Carolina. Bull. Mar. Sci. Gulf Carib. 55:116-146.

Williams, K. T., and R. R. Whetstone (1940) Arsenic distribution in soil and its presence in certain plants. U.S. Dept. Agr. Tech. Bull. No. 732.

Williams, L. L. (1941) The anti-malaria program in North America, pp. 365-370. *In* A Symposium on Human Malaria. Publ. No. 15. Washington, D.C.: American Association for the Advancement of Science.

Wisseman, C. (1973) Report on the control of lice and louse-borne diseases. Bull. PAHO 7(3):81-85.

World Health Organization (1969) International Health Regulations. Adopted by the Twenty-Second World Health Assembly. First annotated edition 1971. Geneva.

World Health Organization (1971) The place of DDT in operations against malaria and other vector-borne diseases. Official Records of WHO No. 190. pp. 176-182.

World Health Organization (1973) Conference on the safety of biological agents for arthropod control. WHO mimeo. WHO/VBC/73.445. 115 pp.

Wright, J. W. (1971) The WHO programme for the evaluation and testing of new insecticides. Bull. Wld. Hlth. Org. 44:11-12.

3

PUBLIC HEALTH ARTHROPODS AS INTERNATIONAL PROBLEMS

VECTOR CONTROL OPERATIONS OVERSEAS BY U.S. AGENCIES

This section presents a historical review followed by a selective description of the most important programs currently in force, along with some analysis of the decision-making process whereby programs are authorized, pesticides chosen, and changes in policy and procedure formulated and ratified.

Historical Review

From its inception as a nation until the end of the nineteenth century, the attention of the United States was directed inward for the most part, and essentially all of the energies of the country were devoted to the winning and consolidation of the continental land mass.

At the turn of the century, the United States became a colonial nation, largely as a result of the war with Spain and the consequent acquisition of territories in the Caribbean and in Asia that had previously been under the rule of Spain. For the first time, the United States had to face the problems of disease control in tropical areas as a matter of national responsibility, and at the same time it was still concerned with several of these, notably malaria, yellow fever, and dengue, within its own boundaries.

As might be expected, the first federal units to deal with vector control problems outside the United States were the military services. The work of Gorgas in controlling yellow fever in Cuba and Panama, following the investigations of the Walter Reed Commission, and the work of Siler and his associates in the Philippines both provided dramatic examples of the possibilities of protecting man from various tropical diseases by attacking the disease vectors.

Russell (1955) has presented a most interesting and exhaustive review of the early activities of private and public U.S. agencies in the control of malaria. The most extensive programs undertaken were those of Gorgas in Havana and, later, in the Panama Canal Zone (these programs are also discussed in considerable detail in Bayne-Jones 1968).

One interesting element in the Panama story that is often overlooked is that of cost-benefit. Gorgas estimated that the sanitary efforts in the Canal Zone saved $80 million more than they cost during the construction period (Russell 1955). This does not even consider the strong possibility that, without disease control measures in the Zone, and given the morbidity and mortality pattern encountered by the French in Panama, public opinion in the United States might well have caused abandonment of the project. To the present day, environmental and chemical means are still being employed by health authorities of the Panama Canal Company to maintain the essentially malaria- and yellow-fever-free status of the Zone and adjacent portions of the Republic of Panama. Two disturbing developments in Panama are the recent appearances of yellow fever in monkeys not far to the west of the Zone and the detection of drug-resistant falciparum malaria adjacent to the Zone.

Based on the work of Gorgas and Reed in the New World, U.S. Army Medical Service personnel conducted studies on dengue in the Philippines. Attacks on the vector *Aedes aegypti* in portions of the Philippines, chiefly in and around Manila, were somewhat successful, but never to the same degree as in Panama. In addition, extensive drainage works were constructed in the permanent U.S. bases in the Philippines, some of which are still extant.

From around 1910 until World War II there was little change in the overseas operations, and essentially nothing in the way of extensive overseas operations was conducted by U.S. government agencies. In the 1930s, the most striking accomplishments in control of disease vectors outside the United States were those of the Rockefeller Foundation; government programs were almost nonexistent. At the same time, within the United States considerable strides were being made in the elimination of malaria by the USPHS, the TVA, and various state agencies. The strongest emphasis on mosquito control during this period was on such factors as environmental modification, drainage, water level management, and impoundment, with relatively little use of pesticides.

After a long period of relative isolation, the United States abruptly found itself in World War II facing vector-borne disease problems of unprecedented magnitude. At home, the military and USPHS program for control of malaria in war areas laid the foundation for the final eradication of the disease. In the Pacific area, however, malaria and other vector-borne diseases often caused higher losses than enemy action. The achievements in this period are summarized in a number of excellent documents dealing with the history of the U.S. Army Medical Department (Coates 1963, 1964).

During and immediately after the war, the military conducted large-scale programs for the control of mosquitoes and other vectors in many parts of the world, and the technical organizations to accomplish this were for the first time made a permanent part of the military organizations. In addition to mosquito control, operations were conducted against flies, chiggers (China-Burma-India area), phlebotomine sand flies (Mediterranean area), and rodents. While personal protective measures were of obvious interest in combat areas, engineering and chemical methods were also employed. A particularly detailed and thorough account of such activities is given for Guadalcanal and other Pacific islands (Coates 1963) In Europe several louse-borne typhus outbreaks were arrested with DDT. In fact, the development of DDT as a residual pesticide stemmed directly from wartime exploitation of a compound first synthesized many years previously.

At the end of the war some vector and disease control activities were continued by the military forces overseas as part of military government activities, but the emphasis shifted rapidly to the relief and reconstruction efforts of such civilian agencies as the United Nations Relief and Rehabilitation Agency (UNRRA) and later to the Economic Cooperation Administration (ECA). There was a very rapid demobilization of the armed forces, but at least the vector-borne disease control apparatus was not dismantled, as it was after World War I.

After World War II, for the first time, the nation entered into a large-scale aid effort, directed toward friends and former foes alike, and eventually extending to much of the world. Previous relief and assistance efforts, such as those organized by Herbert Hoover after World War I, had been of short duration and of limited scale compared to the post-World War II efforts. The latter were conducted by a number of agencies, generally operating under the aegis of the Department of State and represented at present by the Agency for International Development.

At present two cabinet departments, the Department of Defense and the Department of State (through AID), are concerned with the overwhelming proportion of vector control activities outside the United States, either as direct operators of vector control systems or as advisors and suppliers of technical and monetary assistance. Primary attention in this section will be devoted to these two agencies, with some minor exploration of the roles of the Department of Health, Education, and Welfare (through the U.S. Public Health Service), Department of Agriculture, Environmental Protection Agency, and other federal organizations.

Department of Defense

Since shortly after World War II, the Army, Navy, and Air Force have had essentially similar organizational structures for dealing with vector and pest control problems. The general guidelines, operational policies, and procedures are outlined in the *Military Entomology Operational Handbook* (Anon. 1965). In each service, commanding officers are responsible for enforcement of regulations concerning vector control. Medical personnel are responsible for technical guidance, surveys, and determination of the adequacy of control measures. Engineering or public works personnel are responsible for actual conduct of most control procedures. In time of war, vector control operations in combat zones are conducted by special preventive medicine units organized by the medical department.

In practice, almost all vector control operations in the United States and overseas since World War II have been performed by engineering or commercial contract personnel, with technical advice and inspection by entomologists of the medical services. Overall technical advice is furnished by the Armed Forces Pest Control Board (AFPCB), consisting of representatives of the three armed services, with agency and liaison members from the military and other governmental organizations with interests in vector or pest control (as listed in Table 3-1). The Board and its subcommittees meet several times a year. At least one of these meetings includes extensive reports from the U.S. Department of Agriculture (USDA) Laboratory of Insects Affecting Man and Animals at Gainesville, Florida. This laboratory and its research effort are discussed below.

The AFPCB advises the Department of Defense (DoD) in matters of vector and pest control, and its recommendations

TABLE 3-1 Armed Forces Pest Control Board

Members

- Defense Contract Administration Services
- Defense General Supply Center
- Defense Construction Supply Center
- Defense Personnel Support Center
- Defense Supply Agency
- Office of the Chief of Engineers, Department of the Army
- U.S. Army Medical R&D Command
- U.S. Army Research Office
- Office of the Surgeon General, Department of the Army Health and Environment Division
- Chemical Process Laboratory, Wilmington, Delaware
- Naval Ship Systems Command
- Storage Branch, Naval Supply Headquarters
- Naval Facilities Engineering Command
- Preventive Medicine Division, Bureau of Medicine Department of the Navy
- Biological Sciences Division, Office of Naval Research
- 4500 Air Base Wing, Tactical Air Command
- USAF School of Aerospace Medicine
- Directorate of Civil Engineering, Department of the Air Force
- Office of the Surgeon General, U.S. Air Force
- Facilities Management Division, Headquarters, Marine Corps

Agency Representatives

- U.S. Army General Material and Parts Center
- U.S. Army Environmental Hygiene Agency
- Headquarters, Continental Army Command
- U.S. Army Medical Bioengineering R&D Laboratory
- U.S. Army Mobility Equipment Command
- U.S. Army Natick Laboratories
- Navy Disease Vector Ecology and Control Center, Jacksonville
- Navy Subsistence Office
- Military Sealift Command
- Naval Air Systems Command
- Naval Medical Field Research Laboratory
- Headquarters, San Antonio Air Material Area
- Sanitary Engineer Officer, Headquarters, U.S. Coast Guard

TABLE 3-1 (continued)

Air Force Office of Scientific Research
U.S. Army Academy of Health Sciences
U.S. Army Health Services Command
U.S. Army Material Command

Liaison Members

Federal Working Group on Pest Management
Canadian Defense Liaison Staff
Bureau of Sport Fisheries and Wildlife, USDI
Canadian Forces Headquarters
Office of Health, Agency for International Development
Agricultural Research Service, USDA
Animal and Plant Health Inspection Service, USDA
National Science Foundation
National Research Council
British Liaison Office, Office of the Surgeon General, Department of the Army
Smithsonian Institution
Technical Development Laboratories, Center for Disease Control
National Marine Fisheries Service, Department of Commerce
Office of Environmental Affairs, Department of Commerce
U.S. Environmental Protection Agency
Tennessee Valley Authority

---

are generally accepted by all of the armed services. There are some variations in the patterns of pesticide usage by the individual services, but these are relatively minor. New problems, such as the presence of potentially plague-infected rodents in cargo being returned from Vietnam, are generally addressed by the AFPCB rapidly, and recommendation for action is formulated in a short time.

The AFPCB also maintains the Military Entomology Information Service and receives data on pesticide usage from units of the DoD in the field. In preparing this section of the report an attempt was made to determine the actual amount of pesticides employed outside the United States by Defense Department agencies. At present, data on this subject are handled by Navy computer facilities. According to members of the Armed Forces Pest Control Board (particularly Lt. Cdr. H. W. Fowler, Executive Secretary, and H. G. Russell, Chief Entomologist, U.S. Army Corps of Engineers), program planning data on use of pesticides in various programs are available, but only in the format in which plans for pesticide use are supplied by the agency to the Federal Working Group on Pest Management (FWGPM--see discussion in Executive Committee Report, Chapter 7) for review. It is not readily possible at present to obtain figures for total amounts of each type of pesticide used by major commands or geographical areas. No computer program has yet been written for such a purpose. However, it is planned to do so in the near future, largely to satisfy the needs of a new program in the Army Environmental Health Agency for monitoring military pesticide use. It is anticipated that such data will be available shortly.

It would be possible to obtain *procurement* data on pesticides from the Defense Supply Agency and other procurement agencies. However, this would not provide *use* data, and local commands in the United States and overseas would present the additional problem of separating public health applications from such applications as termite or grain beetle control.

At present, most military pesticide usage by DoD units overseas takes place at fixed installations, such as Clark Air Force Base in the Philippines and similar installations in the Far East, Panama Canal Zone, and elsewhere. Data on public health pesticide usage in South Vietnam are difficult to analyze, particularly as much application was performed by civilian contractors. The nature of that conflict, however, was such that most vector control efforts were centered around large camps, airfields, cantonment areas, and similar installations. In the past 10 years, and at an accelerating pace in the last 3 or 4 years, large U.S. bases overseas have been closed or transferred to local authorities. It may be anticipated that this trend will continue and that pest or vector control operations will diminish.

Department of State
(Agency for International Development)

Immediately after World War II, malaria control operations were initiated in various countries under the Economic Cooperation Administration (ECA), and later under the Technical Cooperation Administration (TCA), which was formed to carry out President Truman's "Point-Four" objectives. The various organizations that were formed under a series of names to continue this work have been reviewed in some detail by Soper (1961). By 1957 the program of malaria control, then under direction of the International Cooperation Administration (ICA), had expended over $60 million (from 1942 to 1952) in countries outside the United States.

Support for a worldwide malaria eradication program was approved by Congress in 1957, and in the next 3 years ICA expended more than $85 million more, in bilateral programs and in donations to international bodies. From the first, this program has emphasized technical advice and guidance and provision of commodities, with actual day-to-day operations conducted by nationals of the recipient countries. At the height of the program such bilateral aid was furnished to 26 countries.

At present, the malaria program is administered by the Office of Health, Technical Assistance Bureau, Agency for International Development. The Washington office is responsible for developing overall policy and assuring its implementation, and it also furnishes central direction for AID-supported malaria eradication research. At times the office is also involved with the control of other arthropod-borne diseases; for example, it is charged with furnishing funds for the WHO onchocerciasis control project in the Volta River basin.

At the height of operations, AID assigned a large number of specialists in vector control, epidemiology, logistics, administration, and other fields to the cooperating countries. A number of these were USPHS personnel; others were recruited specifically for the assignments.

Soper (1961) summarized the status of the program as of 1961 and listed the status of programs in the various participating countries (Table 3-2). This certainly is the largest such operation undertaken by the United States and, when added to the support given to WHO and other United Nations agencies, clearly indicates the extent of the commitment to eradication of malaria.

The Center for Disease Control (CDC) of the USPHS had long worked in malaria control and in 1970 assumed much of the technical direction of the program under agreements with AID, with the latter agency retaining overall control, including budgetary responsibility. A position paper released by AID indicates that the memorandum of understanding to this effect expired on 30 June, 1973, with central direction remaining with AID, as indicated above. The technical direction has shifted to WHO. From the beginning, U.S. policies in the malaria control and eradication fields have been closely aligned with those of the WHO and related agencies.

TABLE 3-2 Status of Malaria Eradication Programs[a] in Countries with ICA Bilateral Projects as of 1961

Areas and Countries	Year Phase Initiated			Population (thousands)							Transmission Known to Occur but No Organized Program
							Eradication Program in Operation				
	Preparation	Attack	Consolidation	Country Total	Total Malarious Areas	Areas where Eradication Achieved	Consolidation Phase	Attack Phase	Preparatory Phase	Total	
Far East											
Cambodia		57S	60S[b]	4,800	1,000	0	0	1,000	0	0	0
Indonesia	59S	60S	64S	86,900	75,000[c]	0	0	2,663	72,337[d]	75,000	0
Laos	57S	58S	61S	2,000	2,000	0	0	951	1,049	2,000	0
Philippines	52S	54S	59S	23,122	8,000	0	2,824	5,020	0	7,844	156
Taiwan	52	53	58	10,030	6,750	5,232	1,258	260	0	1,518	0
Thailand		54S	58S	21,474	21,474	0	7,600	7,000	6,874	21,474	0
Vietnam		59S	63S	13,400	11,683	0	0		11,683	11,683	0
TOTAL				161,726	125,907	5,232	11,682	16,894	91,943	119,519	156
Near East and South Asia											
Ceylon	57	58	59	9,386	6,131	0	2,535	3,596	0	6,131	0
India		58S	60S	397,540	390,000	0	1,500	388,500	0	390,000	0
Iran	56S	57S	59S	20,500	13,000	0	4,170	6,894	1,931	13,000	0
Jordan		59S	63S	1,527	980[c]	0	550[c]	430	0	980[c]	0
Nepal	59S	60S	65S	8,910	4,171[c]	0	0	0	4,171	4,171	0
TOTAL				437,863	414,282	0	8,755	399,420	6,107	414,282	0
Africa											
Ethiopia	57S	58S	62S	20,000	8,000[c]	0	0	250	0	250	7,750
Liberia	60S	61S	65S	1,250	1,250	0	0	350	0	350	900
Libya	58	59	63	1,250	30	0	0	30	0	30	0
TOTAL				22,500	9,280	0	0	630	0	630	8,650

TABLE 3-2 (Continued)

Areas and Countries	Year Phase Initiated: Prepara-tion	Attack	Consolida-tion	Population (thousands): Country Total	Total Malarious Areas	Areas where Eradication Achieved	Eradication Program in Operation: Consolida-tion Phase	Attack Phase	Preparatory Phase	Total	Transmis-sion Known to Occur but No Organize Program
Americas											
Bolivia	56	58	63	3,350	888	0	0	888	0	888	0
Brazil	58S	60S	64	64,007	33,035	3,171	0	4,923	24,941[d]	29,864	0
Colombia	58	59S	63S	13,823	9,787	0	0	9,787	0	9,787	0
Ecuador	57	58S	62S	4,007	2,346	0	0	2,346	0	2,346	0
Guatemala	57	58S	62S	3,618	1,544	0	0	1,544	0	1,544	0
Haiti[e]	57	60	64	3,888	2,800	0	0	0	2,800[f]	2,800	0
Honduras	56	59	64	1,888	1,347	0	0	1,347	0	1,347	0
Jamaica	57	58	62	1,689	1,348	0	185	1,163	0	1,348	0
Nicaragua	57	59S	63S	1,461	1,307	0	0	1,307	0	1,307	0
Paraguay	56	59	63	1,672	805	0	0	805	0	805	0
TOTAL				99,403	55,207	3,171	185	24,110	27,741	52,036	0
GRAND TOTAL				721,492	604,676	8,403	20,622	441,054	125,791	586,467	8,806

[a] Data taken principally from WHO A13/P&B/15, April, 1960 (draft copy).
[b] S, malaria eradication being developed in progressive stages. Date represents beginning date for this phase.
[c] Provisional estimate.
[d] Includes populations in States where, under staged program, operations have not yet begun.
[e] Attack phase interrupted 1958, planned to resume in 1960.
[f] Plans prepared and approved, but operations not yet begun.

SOURCE: Soper 1961. Reprinted with permission.

At present, AID policy is to support malaria programs for countries that have indicated a commitment to such a program by preparing WHO-approved plans and by committing significant resources to the program. AID assistance will include commodity support and perhaps funding of local costs where appropriate.

It is unlikely that the dollar value of U.S. support for malaria eradication and control in other nations will in the foreseeable future return to the levels of a few years ago. Total figures on dollar support are difficult to derive for earlier periods, although Soper (1961) supplied estimated costs (Table 3-3), from 1960 to the end of the program, which indicated a total foreign exchange cost of approximately $390 million. The assumptions on which the table are based are presented in the original paper.

A more detailed analysis of costs from 1957 to 1972 is presented in Table 3-4 (data from Mr. Edgar Smith, AID), with U.S. loan and grant figures for that period totaling more than $313 million. Another measure of the magnitude of the support to the malaria programs is the amount of DDT furnished to the various programs from July 1, 1957, to January 31, 1973, (Table 3-5).

Until relatively recently malaria control and eradication programs had always been justified primarily or solely on a humanitarian basis. However, recent studies such as those of Cohn (1972) have seriously addressed the question of cost versus benefit and the related question of the cost of control versus cost of eradication. Available data are still far too skimpy to allow much in the way of judgments on these matters, much less on the cost-benefit ratios of environmental measures versus pesticide application. The latest report of the WHO Expert Committee on Malaria (WHO 1971a) suggests that these questions will have to be explored in the near future.

Almost all of the pesticide used during AID support of malaria programs has been DDT. A list of other pesticides procured for various programs from July 1, 1966, to July 1, 1973, is presented in Table 3-6.

As noted above, policy decisions and general direction are furnished by the AID Office of Health, Technical Assistance Bureau, with additional advice from a number of consultants as required and with the overall technical advice of WHO. AID has also supported a number of research programs in malaria, described below. In the field of vector-borne disease control, AID support has been almost entirely limited to malaria programs, except for the recent

TABLE 3-3 Some Estimated Costs (in millions of U.S. dollars) of Malaria Eradication from 1960 to End of Program

World Regions	Population (in millions)	Foreign Exchange Cost	Local Currency Cost	Total Cost
Far East	154.1	82.5	198.0	280.5
Near East and South Asia	517.1	133.6	358.8	492.4
Africa	187.9	117.8	282.7	400.5
Latin America	84.8	50.7	101.4	152.1
Europe	19.3	5.0	12.0	17.0
TOTAL	963.2	389.6	952.9	1,342.5

SOURCE: Soper 1961.

onchocerciasis control support mentioned above and some support of vector biology research in East Africa.

Department of Health, Education, and Welfare (U.S. Public Health Service)

Officers and civilian scientists of the USPHS have been involved in overseas vector control operations since the earliest U.S. involvement in such enterprises. Several commissioned USPHS officers were assigned to work with Gorgas in Panama in 1904. Personnel of the Service were also instrumental in the planning and administration of the Pan American Sanitary Bureau and have been, and continue to be, involved in international health organizations.

On a bilateral basis, USPHS professionals were involved in the early post-World War II malaria control and eradication efforts in Greece, Thailand, Turkey, Iran, and other areas. These efforts were coordinated by the Office of International Health, organized in 1945, and now headed by the Assistant Secretary for International Health, HEW.

The largest activity of USPHS in this field in recent years has been the malaria program conducted for AID by personnel of the CDC in Atlanta. This organization provided central coordination and technical experts who

TABLE 3-4 AID Support to Malaria Eradication Programs 1957-1972 (in thousands of U.S. dollars)

Country	Program Date		U.S. Dollars		U.S.-Owned Local Currency Grant	
	Start	End	Loan	Grant	Only	Total
Cambodia	1957	1968	--	772	--	772
Indonesia	1957	1967	--	37,643	--	37,643
Laos	1957	1962	--	1,052	99	1,151
Philippines	1957	*a*	--	8,804	--	8,804
Taiwan	1957	1963	--	167	283	450
Thailand	1957	1972	--	18,496	--	18,496
South Vietnam	1957	*a*	--	8,605	--	8,605
South Korea	1957	*b*	--	68	--	68
Sri Lanka (Ceylon)	1957	1964	--	1,004	599	1,603
India	1957	1972	263,385	83,269	181,970	291,624
Iran	1957	1965	--	4,446	--	4,446
Jordan	1957	1970	--	1,923	--	1,923
Nepal	1957	*a*	--	6,589	8,066	14,655
Pakistan	1962	*a*	23,663	1,286	24,638	49,587
Iraq	1957	1958	--	31	--	31
Afghanistan	1957	*b*	--	44	--	44
Turkey	1957	*b*	--	200	--	200
Ethiopia	1957	*a*	8,388	5,152	--	13,540
Liberia	1957	1965	--	509	--	509
Libya	1957	1965	--	222	--	222
Tunisia	1957	1972	--	534	--	534
Bolivia	1958	1965	--	--	931	931
Brazil	1958	*a*	11,385	14,155	1,696	27,236
Colombia	1958	1964	--	1,762	--	1,762
Ecuador	1957	*a*	2,594	2,162	24	4,780
Guatemala	1962	1969	2,180	2,911	100	5,191
Honduras	1958	1969	2,650	2,497	--	5,147
Jamaica	1958	1965	--	386	--	386
Nicaragua	1966	1972	3,934	2,125	--	6,059
Paraguay	1968	*a*	1,755	272	--	2,027
Costa Rica	1966	*a*	1,451	--	--	1,451
Haiti	1961	*a*	--	17,255	--	17,255
El Salvador	1966	1971	2,665	29	--	2,694
Peru	1960	1965	--	169	--	169
Mexico	1957	1958	--	27	--	27
Panama	1968	1971	1,421	--	--	1,421
TOTAL			88,471	24,566	218,406	531,443

TABLE 3-4 (continued)

---

[a]Active projects. Unliquified balances as of 12-31-72 were: Philippines, $211,000; South Vietnam, $11,000; Nepal, $103,000; Pakistan, $15,000; Ethiopia, $412,000; Brazil, $5,286,000; Ecuador, $156,000; Paraquay, $145,000; Costa Rica, $149,000; Haiti, $111,000.
[b]Ending dates not readily available.

NOTE: Does not include expenditures prior to 1957. Does not include AID contributions to WHO and PAHO Malaria Eradication Special Accounts.

served as advisors to various national and regional malaria program offices. These activities have been greatly diminished in the last 3 years in the interest of "multilateralization"--that is, the transfer of responsibility for the malaria program to the World Health Organization.

The CDC was also the primary coordinating agency for the *Aedes aegypti* eradication program in the United States from 1964 to 1968. While this was, strictly speaking, an internal effort, it was initiated because of specific commitments made by the United States to other members of PAHO. There was informal coordination with other national programs, exchange of information, and so on, but no formal participation by the United States in extraterritorial control activities.

No other vector-borne control operations are being conducted by agencies of HEW/USPHS at present, although CDC, and indeed most of the organizations mentioned, are engaged in significant research on the control of vectors and vector-borne disease. These are described below.

U.S. Department of Agriculture

Although USDA scientists have made many basic contributions to research on vector control, the Department has generally not been involved directly in vector control operations overseas. The USDA Laboratory of Insects Affecting Man and Animals, the primary research agency in the pesticide field for the Department of Defense, has supplied technical experts on an ad hoc basis for control investigations overseas on a number of occasions.

USDA scientists have also conducted studies on the possibilities of control of tsetse in Africa by use of the

TABLE 3-5 Summary of DDT Purchased for AID Malaria Eradication Programs FY 1958 to FY 1973

	DDT--75%			DDT--100%		
Fiscal Year	Quantity (Lb)	Average Cost per Lb ($)	Cost ($)	Quantity (Lb)	Average Cost per Lb ($)	Cost ($)
1958	29,448,800	0.26	7,656,688	69,700	0.217	15,114
1959	40,481,100	0.26	10,525,086	268,500	0.216	57,996
1960	78,129,950	0.23	17,969,888	910,000	0.235	213,850
1961	68,926,066	0.21	14,474,474			
1962	71,041,490	0.19	13,497,883			
1963	81,413,277	0.17	13,840,257			
1964	49,797,026	0.155	7,718,539			
1965	59,824,400	0.16	9,571,904			
1966	62,234,380	0.18	11,202,188			
1967	41,930,867	0.178	7,463,694	1,139,226	0.16	182,276
1968	57,605,325	0.175	10,080,932	938,400	0.165	154,836
1969	60,191,775	0.173	10,413,177	7,277,195	0.148	1,077,025
1970	37,721,700	0.175	6,503,309	1,862,425	0.16	297,988
1971	29,050,650	0.1778	5,165,206	252,600	0.168	42,437
1972	6,653,150	0.215	1,430,427	191,400	0.205	39,237
1973	14,165,800	0.215	3,045,647			
TOTAL	788,615,756		150,559,299	12,909,446		2,080,770

TABLE 3-6 Procurement of Pesticides Other Than DDT 75 percent and 100 percent FY 1967-FY 1973

Fiscal Year	Country	Product	Strength	Quantity	Cost ($ U.S.)
1967	Nepal	Malathion	50%	32,256 lb	16,773.00
1968	Tunisia	DDT	25% EC	13,750 gal	9,551.00
1968	Jordan	Abate	43% EC	35 gal	1,428.00
1969	Brazil	Malathion	82% EC	1,650 lb	9,818.00
1969	Philippines	DDT	25% EC	49,002 gal	45,817.00
1970	Philippines	Abate	43% EC	750 gal	18,750.00
1970	Philippines	DDT	25%	64,235 gal	47,356.00
1971	Tunisia	Abate	43% EC	215 gal	4,870.00
1971	Tunisia	DDT	25% EC	770 gal	688.00
1971	Philippines	Abate	43% EC	250 gal	5,663.00
1972	None				
1973	Tunisia	Abate	43% EC	375 gal	9,375.00

sterile male technique. In a similar vein, USDA scientists are conducting actual control operations (as well as research) in the southern United States and northern Mexico on sterile male control of the primary screwworm *Cochliomyia hominivorax*. While this is only peripherally a public health problem, it has long-range implications for the supply of protein for human consumption. At present, it is planned to extend the control operations further south in Mexico, with Mexican and U.S. scientists cooperating in an effort to drive the flies south of the isthmus of Tehuantepec. Production and sterilization of the flies for air-drops in the United States and Mexico is done at Mission, Texas, with a Mexican plant to be constructed in the near future according to agreements signed in July 1963.

The reorganization of the various entomological activities of the USDA in 1972 would seem to make it more difficult for the USDA to engage in cooperative overseas research and control operations of public health importance. Another unit of USDA that might be engaged in international operations in the future is the Animal and Plant Health Inspection Service (APHIS), under whose aegis part of the control activities connected with the 1971 epidemic of VEE were conducted in South Texas. At some point in the future a threat from this disease might make it necessary to initiate mosquito control procedures in Mexico. However, no permanent coordinating system with the Mexican government has been set up for this purpose.

Other Government Agencies

The EPA has, under recent legislation, considerable authority over the pest control operations conducted by government and private organizations in the United States. At present, the agency's participation in planning or surveillance of pest control operations outside the United States appears to be minimal.

The Federal Working Group on Pest Management receives and reviews data on proposed pest control operations from federal agencies. As stated earlier, these include Department of Defense projects within and outside the United States. As noted, however, many of the proposals do not distinguish the specific areas in which they are to be used.

VECTOR CONTROL RESEARCH BY U.S. GOVERNMENT AGENCIES

Information on vector control research was assembled from a variety of sources, including interviews with responsible personnel, examination of recent literature, and, most important, the Science Information Exchange of the Smithsonian Institution (SSIE). The reports examined from the last source generally covered the period 1971 to 1973. In relatively few cases did the reports indicate the amount of funds allotted, so this item has been omitted from this report.

Department of Agriculture

This agency has by far the largest involvement in the general field of development and testing of arthropod control measures. Primary effort is directed toward crop and livestock pests, but considerable research is also directed toward the control of disease vectors. In addition to the research conducted within USDA facilities, there are also projects supported in a number of universities, chiefly the land-grant institutions. Some of the most important facilities are listed below:

Agricultural Research Center, Beltsville, Maryland

A number of pesticide programs are conducted at this large research center, including synthesis, screening, and testing of promising commercially developed compounds. It is also the site of pioneering research in pheromones, chemosterilants, microbial agents, and other alternative control methods.

Insects Affecting Man and Animals Research Laboratory, Gainesville, Florida

This is the primary site for research on and development of pesticides of public health importance. This laboratory, originally established at Orlando, Florida, was funded at first from military funds transferred to the USDA. It was the first United States laboratory to work with DDT and has screened thousands of compounds, from USDA or commercial sources for pesticide activity on a wide variety of arthropods of public health importance. It has also had a huge screening program for repellents, and has done fundamental work on attraction and repulsion of mosquitoes. Although now funded directly, the laboratory still has the support of the military as one of its primary missions and works closely with the Armed Forces Pest Control Board and other military agencies. Recently, the laboratory has been very active in development of sterile male release programs for mosquitoes, tsetse, and other insects in cooperation with other government laboratories overseas, both military and civilian. Work is also under way on mosquito pathogens and other alternative control measures. This laboratory is an important national resource which should continue to receive priority in funding.

Entomology Research Division, Lake Charles, Louisiana

This is one of the most active sites in the world for research on pathogens of mosquitoes, which may eventually be used for control programs. Personnel from the laboratory have carried out preliminary field trials in the Far East with a promising nematode parasite of mosquitoes and have developed methods for dispersal of the pathogen.

Screwworm Research, USDA Laboratories at Kerrville and Mission, Texas

This research is directed primarily at a livestock pest, which only very occassionally infests man. However, the basic work on sterilization and release may be applicable to other systems in the future.

Cooperative State Research Service

The Service cooperates with agricultural experiment stations and similar state agencies on a wide variety of projects, mostly in the field of crop and livestock protection, but some public health projects are included. Following is a sampling of projects listed by the Smithsonian SSIE:

New York Control of blackflies; control of insects in homes and public buildings
Kansas Basic studies on arthropods as pests or vectors to man and domestic and wild animals (latest report on face flies)

Arkansas Control of ticks and chiggers in rural communities and recreational areas

New Jersey Mosquitoes in relation to agricultural production and veterinary science; chemical research on insecticides and their formulations

Massachusetts Pathogens and parasites for the control of insects of medical and veterinary importance

Missouri Biology and control of arthropod pests of man and livestock (latest report deals mostly with tabanids such as horseflies)

Texas The role of domestic animals in the epizootology of Venezuelan equine encephalitis

This is only a sampling, for which reports were available at SSIE. There are far more of these projects than can be listed here, but the above are representative of the types of cooperative research in progress.

Department of Defense

Armed Forces Pest Control Board

The AFPCB, mentioned earlier under control activities, also receives reports on research, and directs requests for specific research projects, largely to the USDA Insects Affecting Man and Animals Research Laboratory in Gainesville, Florida. Previously this USDA activity was funded by a transfer of funds from the Defense Department. Even though now funded separately, there is a Memorandum of Understanding between the two agencies that assures their close cooperation. Although the Board is not in itself a research or action agency, it does serve as a clearinghouse for research in vector control of interest to the armed forces.

Department of the Army

Most of the development work on pesticides, dispersal methods, and the like has been conducted for the Army by the USDA laboratory at Gainesville, under the general supervision of the AFPCB. Army personnel have been assigned to the Gainesville laboratory and have cooperated in overseas trials of promising pesticides and methods in Panama, Thailand, Okinawa, and other sites of military interest.

Walter Reed Army Institute of Research, Washington, D.C. This institution has dealt almost entirely with disease transmission studies, with vector control forming a small part of the work at times, usually in cooperation with USDA.

SEATO Medical Research Laboratory, Bangkok. This field operation of Walter Reed Army Institute of Research has concentrated almost entirely on medical and biological studies, with some support given to USDA Gainesville field studies, but with little direct vector control work.

Army Medical Research Unit, Malaya. The unit concentrated almost entirely on medical and ecological studies in recent years. Earlier, it did some work on area control of chiggers and control of scrub typhus.

Medical Equipment Development Laboratory, Ft. Detrick, Maryland. This laboratory has developed a number of pieces of application apparatus, including a cold-fogger for flying insects, a light-weight mass delouser, and several hand-operated sprayers.

Army Environmental Health Agency. Operates the complex system for preliminary toxicological evaluation of promising pesticides submitted, usually, by the USDA Gainesville Laboratory. Some work has also been done on slow-release pesticide formulations.

Army Medical Research and Development Command. The Army, through the Medical Research and Development Command, has an extensive grant and contract program with universities and other groups. Some of these programs are listed below. In recent years an extensive portion of the program has been a search for a systemic repellent, not a vector control procedure directly, but of public health interest.

University of California Development of long-lasting tropical repellents

University of Alberta Oral insect repellents, action of insect taste receptors

State Health Department of New Mexico Development of systemic insecticides and rodenticides for plague control

University of California, Davis Control of blood-feeding and oviposition in mosquitoes

University of Maryland Evaluation of ULV ground equipment for mosquito control

Department of the Navy

Several Navy laboratories, in the United States and overseas, have been engaged in the study of arthropod-borne diseases and to some extent in vector control. Much of this research has been coordinated with the USDA Gainesville laboratory, as has been true of the other military services.

Navy Medical Research Laboratory Number 2, Taipei. This laboratory has been active in Japanese encephalitis research, including recently the control of mosquito vectors, testing of pesticides, and application techniques.

Navy Medical Research Laboratory Number 3, Cairo. Fundamental work has been done here on ticks and tick-borne agents and on sandfly-borne diseases. The lab has also been active in biological studies of arthropods of medical importance, assessment, mapping, and so on and is also working on molluscicides for schistosomiasis control.

Medical Field Research Laboratory, Camp Lejeune, North Carolina. In cooperation with the USDA Gainesville laboratory, this laboratory has been testing promising louse control agents. It has also worked on clothing and skin repellents for chiggers and ticks, tick control, and mosquito distribution and control.

Disease Vector Control Centers, Jacksonville, Florida, and Alameda, California. These organizations have worked in cooperation with the USDA Gainesville laboratory and other government and private agencies in the development and evaluation of pest control techniques, including the ground and aerial dispersal of pesticides.

Other Navy Research Activities. In addition, the Navy sponsors research on vector control and associated problems at a number of universities and other research agencies. A sample of these follows:

University of Hawaii Biological control of annoying insects (latest report deals with *Musca sorbens*)

Northwestern University Control of flies and cockroaches by insect hormones

Cornell University Biological control of muscoid flies

Department of the Air Force

The Air Force has not been as active in vector control research as the Army or Navy. The 5th Epidemiological Flight has operated in the Far East, chiefly in various mosquito studies (e.g., distribution and biology) as a consultant agency in control. The Special Spray Flight, stationed in the United States, has worked with the USDA Gainesville laboratory and other agencies in the development and testing of spraying methods, including ultra-low-volume methods.

Department of State, Agency for International Development

This agency does not conduct vector control research, but several of its units have supported extensive research by cooperating with universities and other agencies.

Office of Health

Because of its interest in malaria eradication, this unit has supported research on many aspects of vector control and associated matters by transfer of funds to the Center for Disease Control, U.S. Public Health Service. These activities are discussed below in the section on that agency. In addition, funds have been provided recently to the University of Illinois for research on biodegradable analogs of DDT. A major effort in recent years has been research on malaria vaccine, not directly a vector control operation, but one that, if successful, would obviate the need for much of the pesticide used in the world today.

Office of Research and Technology

This office has supported basic work on the biology and control of *Aedes aegypti* in its natural habitat in Africa, with the work performed by Notre Dame University. It has also supported successful work on the control of vampire bats and rodent control at the U.S. Department of the Interior Laboratories, Denver. AID has also contributed to the USDA Gainesville laboratory studies of the sterility method of tsetse control in Africa.

Department of Health, Education, and Welfare, U.S. Public Health Service

This is a complex agency, with activities ranging from clinical research at the National Institutes of Health to the Indian Health Service. Vector control interests are spread through the many areas, but two primary units will be discussed here, the National Institutes of Health (NIH), and the Center for Disease Control (CDC).

National Institutes of Health, Bethesda, Maryland

This is the site of a tremendous volume of basic and applied research, but almost no in-house effort is devoted to vector control. This function is carried out through research grants and contracts through the Research Grants Division, and the majority of the products funded come under the purview of the National Institute of Allergy and Infectious Disease (NIAID). For the most part, the grants and contracts administered by NIH are in the area of basic research, with the more applied contracts having been administered by CDC until recently.

The NIAID has selected the field of biology and control of disease vectors as a "special emphasis area" to receive some preference in funding. A sample of the types of research projects supported is given below:

Old Dominion University Sex pheromones in disease-transmitting ticks

Notre Dame University Genetics and reproductive biology of *Aedes*

Johns Hopkins University Biology, genetics, and physiology of disease vectors (*Aedes* mosquito)

State University of Florida, Vero Beach The natural history of mosquitoes

Rutgers University Bionomics of simuliids and their nematode parasites

This is a small sample of a large and complex program of research support, but a full listing of the projects in operation would be beyond the scope of this report.

Center for Disease Control, Atlanta

This is an extremely complex organization, which has recently undergone an extensive rearrangement and a movement of some of its component parts. One of the primary missions of CDC is the surveillance and investigation of disease outbreaks. In the course of such work (such as with plague or encephalitis) the CDC is frequently brought into the vector control field and to some extent into research on vector control. One of its units, the Technical Development Laboratory, Savannah, has been more directly concerned and will be discussed below. One of the most important nonresearch functions of CDC in the field of vector control is the extensive educational program that it conducts for public health officials and technicians.

Technical Development Laboratory. Until recently this unit of CDC was located in Savannah, Georgia, and was supported for several years by a transfer of funds from AID. It served as part of the World Health Organization pesticide evaluation system and also developed equipment, application techniques, pesticide susceptibility techniques, and many other items related to vector control. The unit has now been moved to Atlanta and sharply reduced in manpower and funding.

Ecological Investigations Program, Ft. Collins, Colorado. A number of the activities now concentrated at Ft. Collins (e.g., plague, arbovirus) impinge upon vector control research, although the primary emphasis is on disease investigations.

Central America Research Station, El Salvador. This station was originally founded as a malaria research unit and was funded from AID sources. More recently it has been given a broader disease research mission. Investigations have been undertaken in the past on the ecology and control of *Anopheles albimanus*, and at present station personnel are cooperating with researchers from the USDA Gainesville laboratory in studies on a sterile male release program for that species. This project appears to have considerable promise.

Other CDC Research Activities. In addition to its inhouse research program, CDC has in the past supported on a contract basis a number of research projects that concerned vector control to some degree. This program has been drastically cut, but following is a sample of the type of research supported in 1972-1973:

University of Wisconsin California encephalitis virus studies in Wisconsin (Reduction of breeding sites of *Aedes triseriatus*)

University of California, Los Angeles Cytoplasmic incompatibility for filariasis control

Department of the Interior

Several of the agencies of the Department of the Interior have interests in the field of vector control. Often, these are rather indirect, such as the interest of the Fish and Wildlife Service in the protection of wildlife areas, particularly marshes, from pesticides. The Office of Water Resources Research also contracted with the University of North Carolina to examine the effects of mosquito control ditching on high salt marshes. The Bureau of Sport Fisheries and Wildlife has received AID funds for investigations on the control of bats and rodents.

Other Federal Agencies

One of the most prominent government units in many aspects of vector control research will undoubtedly be the EPA. The Food and Drug Administration is also playing an increasing role in assessment of dangers of pesticides, and there are doubtless a number of other agencies with some degree of interest in vector control--largely through an involvement with pesticides. It is believed, however, that the major U.S. government agencies with interest in vector control research have been identified above.

Conclusions

As noted in the opening paragraph of this section, much of the information on research programs discussed here was derived from the summary sheets provided by the Science Information Exchange of the Smithsonian Institution. Although this service was most useful, it had some deficiencies, and some consideration might be given to the establishment of a more complete reporting system covering research in the field of vector control--and in other fields as well. On interviewing representatives of several federal agencies, it became apparent that not all of the projects in the fields of vector or pest control were represented in the SIE system. Furthermore, the lack of data on level of funding for many of the projects reported in the system makes it difficult to assess the scale of the research operations described.

INTERNATIONAL PROGRAMS FOR VECTOR-BORNE-DISEASE CONTROL

Malaria Eradication Program

Following evidence of the success of residual house-sprays of DDT in eliminating the transmission of malaria by anophelines, a global Malaria Eradication Campaign was inaugurated by WHO in 1956. By 1958 it already included 76 countries. With treatments of DDT at 2 g/m^2, dieldrin at 0.6 g/m^2, or gamma-HCH at 0.5 g/m^2 applied once or twice a year, eradication was expected to be attained after 4 years at an average annual cost of $0.50 per person in the affected area. Insecticides, spraying equipment and vehicles were purchased by UNICEF, planning and consultative services were provided by WHO, and field staff and facilities were provided by each participating country. Since 1957, bilateral aid from the U.S. AID has been extended to a total of 36 countries and at present some form of assistance is given to about half that number. At the peak of the campaign in 1962, some 100 million dwellings were sprayed, housing 575 million people, by 130,000 spraymen; the 50-75 percent water-dispersible formulations employed in that year contained 65,000 (metric) tons of DDT, 4,000 tons of dieldrin, and 50 tons of gamma-HCH.

As a result, by 1961 eradication of malaria was achieved in areas with a population of 370 million, and by 1972 in areas inhabited by 745 million people (i.e., fully 40 percent of the total at-risk population of 1.8 billion). Areas inhabited by an additional population of 292 million had reached the consolidation phase (clean-up by drugs and spot-spraying) on completion of the attack phase and before the posteradication maintenance phase was begun (WHO 1973a). In many countries in the consolidation phase (e.g., Turkey) the number of malaria cases had been reduced by over 99 percent.

At present, malaria transmission has been eradicated from all of Europe, the United States, the Soviet Union, and Taiwan; from nearly all of the Caribbean islands and from parts of Mexico and the Indian subcontinent, and has been nearly eradicated from the Middle East, Argentina, Peru, and Venezuela. However, in 1968 there was a severe recrudescence of *vivax* malaria in Sri Lanka. Also, many parts of Central America remain malarious, as do tropical Africa and much of Southeast Asia; recently, serious recrudescences of malaria have appeared in Pakistan and parts of India. The development of resistance by the anopheline mosquitoes to DDT and dieldrin, and of *Anopheles albimanus* to malathion and propoxur as well, has posed problems. Thus, by 1971 the figure for malaria eradication, which had been increasing by 100 million a year from 1964 to 1966, had risen only to 728 million inhabitants. In that year the insecticide expenditure was approximately 43,000 (metric) tons of DDT, 1,200 tons of malathion, 800 tons of propoxur, 200 tons of gamma-HCH, and 60 tons of dieldrin. The details of the malaria campaign have been reviewed by (among others) Bruce-Chwatt (1970), Lepes (1972), Garcia-Martin (1972), Brown (1973), and Johnson (1973).

At present, there are virtually no malaria-eradication activities being conducted in tropical Africa. The attack phase continues in Afghanistan, Pakistan, and parts of India, and final eradication is being pursued in countries of the Middle East. UNICEF has phased out its support in the eastern hemisphere, and is in the process of phasing out in the Americas. United Nations Development Program (UNDP) has offered replacement support in many countries, while WHO has helped defray the insecticide bill in certain other countries.

Although eradication remains the objective in many regions, in others (particularly tropical Africa) it is now understood that the most that can be expected at present is a steady reduction in the number of malaria cases--in other words, control rather than eradication. Thus, increasing attention is being paid to insecticidal methods other than intradomiciliary residual spraying and to biological and genetic control (Wright et al. 1972). Larval control, by means of organophosphate insecticides such as temephos (Abate), and by larvivorous fish such as *Gambusia*, is part of 20 national antimalaria programs.

Onchocerciasis Control Program

Onchocerciasis, a nematode infection that in its advanced stage causes river blindness, is particularly prevalent in interior West Africa. A program for control of this disease, scheduled to start in 1974 and to continue for 15-20 years at an estimated cost of some $6 million per year, was nearing completion of its planning stages as of this writing. The program is principally based on the proposed elimination by insecticide of larvae of the blackfly vector *Simulium damnosum* from 14,000 km of rivers and streams in the drainage basin of the Volta River and parts of the Niger basin (LeBerre 1974). This is to be a UNDP program financed by contributions collected by the World Bank.

WHO completed 3 years of work on the vector control and epidemiological elements of this program. This involved the testing of fixed-wing aircraft and helicopters to apply the insecticide, the development of suitable formulations for water treatment, and the choice of the insecticide from among the candidate compounds methoxychlor, chlorpyrifos, chlorpyrifos methyl, phoxim, chlorphoxim, and temephos. Annual grants of the order of $100 thousand from both the AID and the German Federal Ministry for Economic Cooperation supported this preparatory work by WHO staff and consultants and allowed the initial pilot project to be carried out in the Comoe-Leraba drainage system of Upper Volta in 1973.

On the basis of tests for effectiveness against *Simulium* conducted at the Organisation des Recherches Scientifiques et Techniques Outre-Mer (ORSTOM) Centre at Bobo Dioulasso, Upper Volta, and for lack of toxicity to nontarget aquatic organisms conducted at the ORSTOM Centre at Fort Lamy, Chad, the larvicide selected for initial use was temephos, to be applied at a water concentration of 0.05 ppm. The principal method of application will be a quick-release delivery from aircraft on target sections of the streams and rivers. The annual requirements for temephos will be a maximum of 40 tons active ingredient in 1977. Previous onchocerciasis control activities had never expended more than 0.5 tons of DDT, at concentrations not less than 0.1 ppm.

Local services are to be provided by the governments of the seven countries involved--Upper Volta, Ivory Coast, Mali, Niger, Ghana, Togo, and Dahomey--where it is estimated that there are more than a million cases of onchocerciasis, about 10 percent of which have caused blindness. The onchocerciasis program will be accompanied by activities to control trypanosomiasis and its tsetse-fly vectors, to ensure the health of the human population and cattle which are expected to recolonize the areas depopulated because of onchocerciasis.

Aedes aegypti Control and Eradication

The attempt to eradicate this yellow fever vector first started in Brazil in 1938 and was extended in 1947 to other South and Central American countries under the aegis of the Pan American Sanitary Bureau (now PAHO, the WHO Regional Office for the Americas). Eradication was successful in Central America and most South American countries, while the Caribbean islands, Venezuela, and the

Guianas remained infested (Severo 1952). The United States joined the eradication operations in 1964, using DDT as the principal insecticide, but discontinued them in 1968. During the last decade, reinfestations have appeared in Central America, Colombia, and northern Brazil, but these are being combated (Schliessmann and Calheiros 1974). The insecticide usage in 1970 in a single country--Venezuela--was 2.2 metric tons: half fenthion, a quarter malathion, and a quarter temephos. It is expected that operations will be continued and intensified against *Aedes aegypti*, since it is a vector not only of yellow fever but also of the dengue viruses (Soper 1972).

The importance of dengue hemorrhagic fever in Southeast Asia has led to the development of measures to control the heavy infestations of *A. aegypti* in the towns and cities. Sand granules containing 1 percent Abate are used to eliminate the larvae in domestic water containers (Bang et al. 1972). To control the adults, great success has been obtained with ULV aerosols of malathion or fenitrothion applied by portable or vehicle-transported machines (Pant et al. 1971); these have a degree of residual effectiveness to reinforce the direct kill (Pant and Mathis 1973), and if the applications are repeated at weekly intervals over a period of 1 month, the control of the *Aedes aegypti* population will last for 4 months thereafter (Pant et al. 1973). ULV airsprays of malathion were also found to be effective (Lofgren et al. 1970). These methods were developed by a WHO research unit in Bangkok, but there is no formal dengue hemorrhagic fever control program.

For the control of yellow fever in Africa, where there are other species of *Stegomyia* vectors besides *Aedes aegypti*, the main reliance is on vaccination programs. Preparations have been made by WHO for emergency response to epidemics by means of ULV aerial sprays (Brooks et al. 1970) and thermal aerosols, and a WHO research unit has been established at Enugu, Nigeria, to study methods of vector surveillance and control. WHO also maintains a worldwide computer-based surveillance system of the distribution and density of *Aedes aegypti* and other *Stegomyia* vectors.

Filariasis and Encephalitis Control

Although it is estimated that 250 million people suffer from this disease, there is no international program for filariasis control like that for malaria. The Government of India started a national program 20 years ago, but the target vector *Culex fatigans* developed resistance to the DDT, dieldrin, and gamma-HCH used as larvicides and residual insecticides, and operations have reverted to the use of larvicidal oils. A WHO research unit in Rangoon has developed a successful control method using periodic applications of fenthion to give a concentration of 1 ppm in the breeding places (Self and Tun 1970, Graham et al. 1972), and this has been adopted by the Government of Burma. In West Africa, chlorpyrifos (Dursban) has been found to be an effective larvicide for this vector, and it lasts longer than Abate (Subra et al. 1969).

At present a research unit in New Delhi, run jointly by WHO and the Indian Council of Medical Research, is investigating the release of chemosterilized males as a means of controlling *Culex fatigans* over large areas; strains with chromosomal translocation are also being tested as a means of genetic control. For biological control of larvae, WHO has investigated guppy fish (*Poecilia reticulata*) in Rangoon and Bangkok and mosquito fish (*Gambusia affinis*) in wells in New Delhi. The possible effectiveness of the nematode *Reesimermis nielseni* against *Culex* larvae has been investigated by WHO in Bangkok and Taipei.

The biology and control of *Culex tritaeniorhynchus*, the vector of the Japanese encephalitis virus, has been investigated by WHO units in Seoul and Taipei (Self et al. 1973). Effective control of adult populations of this vector has been obtained by aerial ULV sprays in Korea and ground ULV sprays in northern Thailand, fenetrothion proving superior to malathion.

Chagas' Disease and Trypanosomiasis Control

It is estimated that 30 million people in South America suffer from the type of trypanosomiasis known as Chagas' disease. To control the vectors, which are triatomine bugs that lurk in the crevices of house walls, residual sprays of dieldrin or gamma-HCH have been employed in many South American countries. Now that resistance is appearing to these organochlorines, these countries would substitute the carbamate insecticide propoxur if they could afford the much greater cost. WHO has inaugurated a Chagas' Disease Vector Control Unit in Acarigua, Venezuela, aided by USPHS grant funds for the first two years of 1973 and 1974.

Control of trypanosomiasis causing nagana disease in livestock and sleeping sickness in man has been carried out with DDT and dieldrin applied as residual sprays to riverine vegetation in West Africa and as fine sprays or aerosols applied from the air to thicket areas in East Africa, against the tsetse vectors. A number of organophosphorus compounds, e.g., fenthion, are effective as residual sprays or ULV aerial sprays, but their expense has so far prevented their adoption.

Other Vector Control Activities

For control of the *Phlebotomus* sandflies that transmit the two main types of human leishmaniasis, DDT still remains highly effective. For control of body lice transmitting epidemic typhus, DDT and gamma-HCH are encountering resistance and are generally being superseded by malathion and carbaryl powders; however, malathion resistance has already appeared in Burundi and parts of Egypt. In areas of flea-transmitted plague, rodent control is often supplemented by DDT or diazinon dusting powders. There are no international programs for these diseases, although WHO is involved in standardizing methods of surveillance, control, and resistance testing.

Program for Evaluating and Testing New Insecticides

This program was started by WHO in 1960, when the challenge of insecticide resistance called for substitute insecticides for use in malaria eradication. The search was subsequently reinforced by the need to minimize environmental contamination. Over 1,400 candidate insecticides have been tested, submitted by 43 manufacturing companies and 5 universities and institutes (Wright 1971).

The compounds pass through a succession of seven stages by a process of trial elimination. Stage I consists of screening tests, performed at Riverside, California, and Urbana, Illinois. In Stage II, practical effectiveness tests are carried out in the laboratory (Gainesville, Florida, and Porton, England), and Stage III consists of simulated field trials (Gainesville and Savannah, Georgia). In Stage IV, true field tests are run on the single-hut scale (Porton; Savannah; Gainesville; Arusha, Tanzania; and Bobo Dioulasso, Upper Volta). Stage V consists of village-scale trials (Kaduna, Nigeria), and Stage VI is a district-wide operational trial (Kisumu, Kenya). In the

Stage VII trial, district-wide epidemiological assessment of malaria is added to the entomological assessment.

In order to avoid carrying to an advanced stage any compounds that are insufficiently free of toxic hazard, their mammalian toxicity is assessed at Stage II (Carshalton, England) and their absolute safety for field use is investigated in the Stage V village trials.

So far, only malathion (an organophosphate) and propoxur (a carbamate compound) have successfully proceeded through all seven stages; fenthion, although it is highly effective, causes cholinesterase reduction in spraymen. Carbaryl, methoxychlor, phenthoate, jodfenphos, bromophos, and phoxim are very good but are insufficiently promising at present to go to the expensive Stage VI and VII trials. Fenitrothion was selected for a Stage VI-VII trial. Chlorphoxim is very promising, as is Landrin, a carbamate compound now being used experimentally in Central America.

This program, while centered on the discovery of residual insecticides for malaria eradication, also encompasses larvicides for anopheline, culicine, and simuliid vectors, the later stages of testing being performed at the WHO field research units.

Use of DDT in Public Health Programs

At the peak of the global malaria eradication campaign in 1962, 65,000 (metric) tons of DDT were used in the program, as well as 4,000 tons of dieldrin and 500 tons of gamma-HCH; in that year one-third of the DDT produced in the United States went to the malaria campaign. At that time, the world production of DDT was between 200,000 and 300,000 metric tons.

By 1969, the public health usage of DDT was 31,500 (metric) tons, and the estimated world DDT production was down to 95,400 tons.* The United States produced about half of that, with approximately 23,600 tons going to public health programs abroad. Another 13,700 tons was exported for agricultural purposes, as a part of the 53,500 tons DDT used in agriculture overseas in that year. At least a dozen other countries were producing DDT.

---

*As noted at several points in this report, accurate figures for world--and domestic--pesticide production and use are extremely difficult to generate. These data are, at best, rough estimates. They are taken from a draft unpublished document prepared for AID by R. von Rumker Consultants.

Between 1962 and 1971, the use of DDT for other public health purposes was considerably less than 1,000 tons annually. Estimated figures in the year 1967 are 60 tons for trypanosomiasis (*Glossina*) and only 0.5 tons for onchocerciasis (*Simulium*), and from 10 to 50 tons for *Aedes aegypti*. At that time the use of DDT against pest mosquitoes in the United States, Caribbean, and Mediterranean areas was rapidly decreasing, but the use of DDT-containing proprietaries in households was still quite high throughout the world (e.g., in Greece and India). By 1972, DDT had been almost completely replaced by organophosphorus compounds as mosquito larvicides.

Safe Use of Pesticides

WHO has been carrying out clinical studies in Brazil and India on large groups of spraymen with a history of at least 7 years of applying DDT; these studies have so far failed to show any consistent difference in the condition of such groups and control groups without such exposure. A release from the *Official Records* of WHO (1971b) states that there is no sound reason to believe that the millions of people protected by DDT against vector-borne diseases are at tangible risk from their small exposure to the chemical.

It has been estimated that there are about 500,000 cases of insecticide poisoning annually, with not less than 1 percent mortality (WHO 1973b). Because of the extent of this problem, WHO is cooperating with FAO in organizing regional seminars to bring together the staff of agriculture and health ministries. UNDP plant-protection projects in several countries offer other opportunities for WHO to cooperate with FAO in promoting the safe use of pesticides. WHO is issuing data sheets on the toxicity of individual insecticides and their treatment and is developing a procedure for the regular and uniform reporting of cases of insecticide poisoning. It has developed routines for emergency action in case of epidemics of poisoning (e.g., the eating of seeds treated with mercurial fungicides).

Other International Programs

The Danish International Development Agency supports annual international training courses in pesticide chemistry and in vector and rodent control. The Swedish equivalent has given grants for studies in vector ecology. The Netherlands government maintains a tropical disease institute in Nairobi and contributes to the UNDP-supported International Centre for Insect Physiology and Ecology in that city. The government of the Federal Republic of Germany has assisted with grants for developmental research and has supplied insecticides. The government of the Soviet Union has conducted, as part of its UNDP contribution, international courses in vector ecology and vector control. The Overseas Development Administration of the United Kingdom, through its Centre of Overseas Pest Research, participates in a number of pest-control projects throughout the world. The International Development Research Centre of the Canadian government supports research on biological control of *Simulium* in Africa. The French Organisation des Recherches Scientifiques et Techniques Outre-Mer has scientific missions established for vector control in several African countries. Many UNDP programs for agricultural development, especially in irrigated areas, have a public health component. Larger public health projects, e.g., schistosomiasis control in upper Egypt and onchocerciasis control in West Africa, have been submitted to the World Bank for financing.

REFERENCES

Anonymous (1965) Military Entomology Operational Handbook. Washington, D.C.: U.S. Government Printing Office.

Bang, Y. H., N. G. Gratz, and C. P. Pant (1972) Suppression of a field population of *Aedes aegypti* by malathion thermal fogs and Abate larvicides. Bull. WHO 46: 554-555.

Bayne-Jones, S. (1968) The Evolution of Preventive Medicine in the United States Army, 1607-1939. Washington, D.C.: U.S. Government Printing Office.

Brown, A. W. A. (1973) Pest control strategies ten years hence: malaria. Bull. Entomol. Soc. Am. 19:193-96.

Brooks, C. D., P. Neri, N. G. Gratz, and D. B. Weathers (1970) Preliminary studies on the use of ultra-low-volume application of malathion for control of *Aedes simpsoni*. Bull. WHO 42:37-54.

Bruce-Chwatt, L. J. (1970) Global review of malaria control and eradication by attack on the vector. Misc. Publ. Entomol. Soc. Am. 7:17-23.

Coates, J. B. (ed.) (1963) Medical Department, United States Army, Preventive Medicine in World War II. Vol. VI--Communicable Diseases, Malaria. Washington, D.C.: U.S. Government Printing Office.

Coates, J. B. (ed.) (1964) Medical Department, United States Army, Preventive Medicine in World War II. Vol. VII--Communicable Diseases, Arthropod borne Diseases Other than Malaria. Washington, D.C.: U.S. Government Printing Office.

Cohn, E. J. (1972) Assessment of malaria eradication costs and benefits. Am. J. Trop. Med. Hyg. 21(5):663-667.

Garcia-Martin, G. (1972) Status of malaria eradication in the Americas. Am. J. Trop. Med. Hyg. 21:617-633.

Graham, J. E., M. H. M. Abdulcader, H. L. Mathis, L. S. Self, and A. Sebastian (1972) Studies on the control of *Culex pipiens fatigans* Wiedemann. Mosq. News 32: 399-416.

Johnson, D. R. (1973) Recent developments in mosquito-borne diseases: Malaria. Mosq. News 33:341-347.

LeBerre, R. (1974) *Simulium damnosum*, pp. 55-63. *In* Control of Arthropods of Medical and Veterinary Importance, R. Pal and R. H. Wharton (ed.). New York: Plenum Press.

Lepes, T. (1972) Research related to malaria: a review of achievements and further needs. Am. J. Trop. Med. Hyg. 21:640-647.

Lofgren, C. S., H. R. Ford, R. J. Tonn, Y. H. Bang, and P. Siribodhi (1970) The effectiveness of ultra-low-volume applications of malathion at a rate of 3 fluid ounces per acre in controlling *Aedes aegypti* in Thailand. Bull. WHO 42:27-35.

Pant, C. P., and H. L. Mathis (1973) Residual effectiveness of ULV aerosols against *Aedes aegypti* in Bangkok: A study of malathion and fenitrothion applied by a portable ULV machine. Southeast Asian J. Trop. Med. Public Health 4:231-237.

Pant, C. P., G. A. Mount, S. Jatanasen, and H. L. Mathis (1971) Ultra-low-volume ground aerosols of technical malathion for the control of *Aedes aegypti*. Bull. WHO 45:805-817.

Pant, C. P., M. J. Nelson, and H. L. Mathis (1973) Sequential application of ultra-low-volume aerosols of fenitrothion for sustained control of *Aedes aegypti*. Bull. WHO 48:455-459.

Russell, P. F. (1955) Man's Mastery of Malaria. London: Oxford University Press.

Schliessmann, D. J., and L. B. Calheiros (1974) A review of the status of yellow fever and *Aedes aegypti* eradication programs in the Americas. Mosq. News 34:1-9.

Self, L. S., and M. M. Tun (1970) Summary of field trials in 1964-69 in Rangoon, Burma of organophosphorus larvicides and oils against *Culex pipiens fatigans* larvae in polluted water. Bull. WHO 43:841-846.

Self, L. S., H. K. Shin, K. H. Kim, K. W. Lee, C. Y. Chow, and H. K. Hong (1973) Ecological studies on *Culex tritaeniorhynchus* as a vector of Japanese encephalitis. Bull. WHO 49:41-47.

Severo, O. P. (1952) Eradication of the *Aedes aegypti* mosquito from the Americas. Mosq. News 16:115-121.

Soper, F. L. (Chairman) (1961) Report and Documentations on Malaria. A Summary. Am. J. Trop. Med. Hyg. 10(4): 451-502.

Soper, F. L. (1972) International health--2000 A.D. Bol. Ofic. Sanit. Panam. 72:397-408.

Subra, R., B. Bouchite, and J. Coz (1969) Evaluation sur le terrain de l'efficacité de deux insecticides organophosphorés, l'Abate et la Dursban, contre les larves de *Culex pipiens fatigans*. Med. Trop. (Marseille) 29: 607-614.

World Health Organization (1971a) WHO Expert Committee on Malaria, Fifteenth Report. WHO Tech. Ser. No. 467.

World Health Organization (1971b) The place of DDT in operations against malaria and other vector-borne diseases. Offic. Rec. WHO No. 190. pp. 176-182.

World Health Organization (1973a) Malaria eradication and other antimalaria activities in 1972. WHO Chron. 27: 516-524.

World Health Organization (1973b) Safe Use of Pesticides: 20th Report of the WHO Expert Committee on Insecticides. Who Tech. Rep. Ser. No. 513, p. 42.

Wright, J. W. (1971) The WHO programme for the evaluation and testing of new insecticides. Bull. WHO 44:11-22.

Wright, J. W., R. F. Fritz, and J. Haworth (1972) Changing concepts of vector control in malaria eradication. Annu. Rev. Entomol. 17:75-102.

4

ALTERNATIVE TACTICS AND STRATEGIES

VECTOR AND PEST CONTROL BY ENVIRONMENTAL SANITATION

Before the advent of DDT and the other residual insecticides, the basic principle of malaria vector control was "species sanitation." Within the problem area the vector was identified, its breeding habitats were delineated, and control measures were designed to eliminate or reduce the larval population, either by repeated application of larvicides or by elimination of the breeding places through drainage or other engineering methods. Clearly, this is an "alternative" method quite different from the novel and experimental ones mentioned later in this section, in that it is already a widespread and accepted method of control. Early mosquito control programs depended heavily on methods of environmental sanitation. With the introduction of organic insecticides these methods dropped into the background in many areas. Recently, with increased problems of insect resistance and environmental contamination caused by the use of insecticides, "source reduction" is of increasing importance. It is recognized, too, that careful research on the biology and behavior of the pest species in the field will be necessary as a basis for the development of increasingly sophisticated forms of environmental manipulation.

There have been innumerable programs for control of arthropod diseases based on this type of source reduction in the past. Urban yellow fever was banished from tropical American cities by destroying or removing the artificial water containers in and around houses. Clearing of vegetation in strategic areas frequented by man has been effective in the control of African sleeping sickness. Proper handling and disposal of garbage and wastes from homes, restaurants, and industrial plants, adequate sewerage systems, and removal of stabled animals has reduced the former scourge of flies from urban environments. It is obvious that it is impossible to summarize and arrive at a cost-benefit analysis of even a small portion of these control measures, and so this section is restricted to discussion of several major mosquito control projects about which it has been possible to obtain some information.

The Panama Canal Zone

Panama's former reputation for high morbidity and death rates is well founded. Because we now take for granted the generally healthful living conditions in Panama and elsewhere in the tropics, it is appropriate to remind ourselves of what the situation was before disease vector control. Robinson (1907) estimated that during the construction of the Panama Railroad from 1850 to 1855, although "not more than forty percent" of the employees died in service, the hospitals were always filled with patients suffering from chills and fever. From the inadequate mortality statistics available, Gorgas estimated that from 1883 through 1904 the death rates in Panama City ranged from 5 to 31 per 1,000 population per annum. In 1885, of a total population of 20,276 there were 687 deaths attributed to malaria. It was estimated that over 75 percent of all hospital patients had malaria.

The failure of the French Canal Companies because of high death rates caused by malaria and yellow fever is well known. Chamberlain (1929) said, "it has been estimated that at least 16,000 employees died from all causes during nine years, this mortality occurring in a total force which did not average over 10,121 for the period."

In 1904, Gorgas began his mosquito control campaign in Panama. It was limited to certain areas within the Canal Zone, and to Panama City and Colon at each end of the canal. The attack on yellow fever consisted of quarantine against importation of new cases, isolation of patients, use of screening and bed nets, fumigation of all buildings, and destruction of the breeding places of *Aedes aegypti*. It is impossible to estimate how much source reduction contributed to the elimination of human yellow fever cases by May 1906. However, Simmons (1939) states that fumigation alone did not prevent an increase in the disease among nonimmune new arrivals.

According to LePrince, who was responsible for the early antimalaria program, malaria was being transmitted among patients in the unscreened French Canal Hospital itself. Malaria control was accomplished by a combination of several methods: use of quinine; screening and bed nets; killing of adult mosquitoes in houses and removal of sheltering underbrush; use of oil and other larvicides; source reduction by filling, drainage, water level management, and redirection of streams. This included the installation of permanent subsoil tiling and concrete-bottomed ditches. Again, all of these methods contributed to the dramatic reduction in malaria in Panama City and Colon, but the importance of sanitary measures is reflected in the statement by Simmons: "At present [1939], little use is made of quinine as a prophylactic, the killing of adult mosquitoes is no longer a routine control procedure, and few houses in these two cities are screened. However, both Panama City and Colon are well drained, and are relatively free from anopheline breeding places."

Simmons (1939) refers to an estimate of the potential malaria rate in the unsanitized Canal Zone in 1904 as 3,000 per 1,000, or three attacks per individual per year. In 1907 the malaria admission rate was 424 per 1,000, and in 1916 it was down to 16 per 1,000, where it was more or less stabilized.

Following the preliminary control measures, the main reliance for mosquito control, including pests as well as malaria vectors, has been elimination of breeding places by drainage and filling. Although there have been considerable expenditures for permanent tiled and concrete-lined ditches, there also are hundreds of miles of unpaved earthen ditches. With periodic cleaning, these have produced excellent results. Subsoil tile drains have been installed in some swampy areas. In the vast sea-level swamps, earthen intercepting ditches have been dug so as to bring about quick drainage of rainwater and an influx of seawater. The sanitation methods have been supplemented by the use of larvicidal oil, which is applied to those areas that cannot be drained. In large areas inaccessible by land, Paris green has been distributed by airplanes.

In 1945, DDT was introduced as a larvicide; it was applied as 10 percent dust to aquatic vegetation in the Chagres River. An average of 5,000 pounds of DDT was used each year. In 1945 5 percent DDT was also used as a residual spray for malaria eradication. In 1957 and 1962, 1 percent dieldrin was used for this purpose.

It appears to be impossible to estimate how much money has been saved by permanent drainage systems and other measures of mosquito control, that otherwise would have been spent in repetitive insecticidal applications. According to information received from Mr. J. P. MacLaren,

Chief, Division of Sanitation of the Canal Zone (Personal communication, 1973), there is continuous maintenance of more than 500 miles of ditches. Mr. Charles M. Keenan, Environmental Health Division (Personal communication, 1973) says that drainage maintenance continues to be the major method of control, representing about 85 percent of the total effort. Townsites are inspected for larvae and are routinely fogged with fenthion. No additional engineering projects for malaria control are being contemplated because, as Mr. MacLaren says, "Our insect-borne diseases are under control."

From 1935 through 1939, the Republic of Panama spent a total of $330,830 for malaria control, or an average of $66,146 per year (Kumm 1941). Drainage systems have been installed in about 40 towns along the Central Highway (C. M. Gallardo, Personal communication, 1973). Such permanent measures are effective in urban communities but are not economically feasible for the scattered rural populations. The Republic began the attack phase of its malaria eradication program in 1957, by spraying all houses with 5 percent DDT. This required the application of about 200,000 pounds of DDT every 6 months. According to the XVIII Report of the Pan American Sanitary Conference, the eradication program has a yearly budget of $647,000. There was a steady decline in malaria in the Republic, and from 1966 through 1971 the annual number of reported cases ranged from 1,041 to 3,726. Only a small proportion of the reduction in the incidence of the disease can be attributed to the local drainage systems referred to above; however, according to Gallardo, they did bring about a reduction in incidence of malaria from 60 percent in 1930 to less than 5 percent in 1946 in the communities in which they had been installed.

El Salvador

One of the most serious problems of persistent malaria is found in El Salvador. According to Kumm (1941), from 1935 to 1938, drainage systems were installed in three localities, and more than 4,000 inverts and 25,000 lengths of concrete pipe for subsoil drainage were manufactured. This program cost $1,265 annually, about 1 percent of the total budget of the Health Department.

A drainage project was built in San Miguel in 1934, sponsored by the Rockefeller Foundation. Valdivieso (1950) states that, following the establishment in 1942 of a collaborative project (SCISP) of the General Health

Services and the U.S. Government, this system was completed with 7,947 linear meters of paved ditches, 925 meters of underground drains, and 525 meters of dirt ditches, which protected a total of 19,000 persons. Mr. George R. Whitten, an agricultural engineer, inspected this project in 1969, and comments, "This is source reduction in its simplest and most classical form . . . . The City of San Miguel is responsible for maintenance on the system and obviously they take this responsibility for maintenance seriously. Signs of recent replacement of sections of the lining is evident, dirt and trash is constantly removed from the bottom . . . . I would say that the system is working just as well now as it was 35 years ago when it was constructed."

A second drainage project inspected in 1969 by Mr. Whitten was installed in 1944 in Sonsonate. This is a joint irrigation and drainage project in which excess irrigation water was conveyed through drainage ditches back into the irrigation system, with the result that a large area on the outskirts of Sonsonate was converted from an *Anopheles albimanus* breeding area to the production of cotton, pasture, okra, and corn. Mr. Whitten stated that the system "is still in excellent condition accomplishing the task assigned." He says further, "It would be very interesting to compare the initial cost and maintenance over the last twenty-nine years with the cost of insecticides and labor necessary to control the breeding in the area." One should also add to this equation the economic gain resulting from changing mosquito-producing swampy areas into highly productive agricultural land. Unfortunately, we cannot evaluate the impact of this program on malaria reduction, as its effects would have been obscured by those of the residual DDT malaria eradication measures.

Nave Rebollo et al. (1973) divide the antimalaria activities in El Salvador into three phases. The first was begun in 1932 and consisted of installing drainage systems in eight important localities. The second phase began in 1946 with the application of DDT to houses. In 1955 the country entered into an all-out eradication program; this was the third phase. It is of interest to compare the success of these three phases.

In the period from 1932 through 1944, source reduction was supplemented by the use of larvicides. Statistics compiled from 1930 through 1945 show no reduction in deaths caused by malaria throughout the country: in 1930 the number of malaria deaths per 100,000 people was 217; in 1944 it was 215. Table 4-1 shows infection rates in

TABLE 4-1 Rate of Malaria Infection in Five Towns in El Salvador with Drainage Systems

	% of Population Found Positive for Malaria			
	1938		1949	
Town	Enlarged Spleen[a]	Parasite Detected[b]	Enlarged Spleen	Parasite Detected
Atiquizaya	31.7	2.0	32.4	3.6
Santa Ana	13.8	2.4	22.4	0.8
Sonsonate	16.4	1.4	10.8	2.4
Usulutan	26.7	2.2	11.8	2.5
San Miguel	30.6	6.0	14.1	1.7

[a]Indicates previous infection.
[b]Indicates current infection.
SOURCE: Nave Rebollo et al. 1973.

five of the sanitated communities. It is evident that at least in Sonsonate, Usulutan, and San Miguel there was a significant diminution in malaria prevalence. However, it is difficult to assess the contributions of drainage alone in this reduction, as larvicides also were applied. It should be noted, however, that up to the time the 1949 survey had been made, these towns had not been treated with DDT.

The eradication program has reduced malaria mortality in El Salvador--to only 3.4 and 2.6 deaths per 100,000 people recorded for 1970 and 1971. As is well known, the disease has not been eradicated and in fact, when one compares the annual parasite indices from 1956 through 1964 with those of the years 1965 through 1971, there appears to have been an increase in the number of infections. Nave Rebollo et al. conclude that the partial spray with DDT and OMS-33* appears incapable at present of interrupting transmission; this is also the case with partial collective treatment which is also unable to keep the morbidity curve stable.

The total cost of the eradication program from 1956 through 1972 was 45,415,832 colones, or over $18 million. If eradication cannot be achieved, a never-ending program of insecticide application and drug administration will be necessary. Although the results of the drainage systems

*Propoxur (Baygon).

were not spectacular, it seems obvious that where they can be installed they will be a valuable supplementary measure, and in fact may become the principal means of control for certain communities. Drainage schemes need not always be elaborate or expensive. An example is given by Carmichael (1972), who considers the *esteros*, the coastal mosquito breeding area, to be the "most serious contribution to malaria in El Salvador." These are accumulations of fresh water that are separated from the beach by sand barriers created by wave action. Mr. Carmichael stated that the esteros could be opened to the sea if ditches were dug through the sand barriers in proper timing with the tide cycles.

Tennessee Valley Authority

The hydroelectric program of the TVA has converted almost the entire drainage system of the Tennessee River into a series of man-made impoundments. In the southeastern United States, the malaria vector is *A. quadrimaculatus*, which breeds in quiet, pooled waters, and the innumerable breeding areas created along the vast shoreline represented a health hazard of considerable magnitude. The 24 impoundments cover more than 600,000 acres and have over 10,000 miles of shoreline. Gartrell et al. (1972) refer to earlier malaria surveys that showed that in northern Alabama more than a third of the people living within a mile of the planned reservoir margins were infected with malaria. Hence, control of the malaria vector was incorporated in the original planning of the reservoirs.

The following are the main steps in the management of *A. quadrimaculatus* breeding in the TVA reservoirs:

1. Reservoir preparation: removal of vegetation and debris from the areas to be flooded.
2. Water level management to strand the aquatic stages or to flush them into open waters where they become more available to predators. Ideally, this would involve a flood surcharge prior to the breeding season, followed by a drawdown to strand accumulated flotage; cyclical water level fluctuation during the early part of the breeding season; and cyclical fluctuation accompanied by a gradual drawdown during the latter part of the season in order to keep the mosquito-producing areas below the encroaching marginal vegetation.
3. Routine maintenance of marginal drainage systems.

4. Pumping water from low areas behind dikes; operation of diked areas as lateral impoundages.
5. Removal of regrowth of vegetation in order to maintain a clean shoreline.

An appraisal of the effectiveness of this program was given in 1941 by Hinman:

> In 1935, owing to construction requirements at Wheeler Dam, it was essential that the lake be held constantly at maximum elevation throughout the mosquito breeding season. Despite the fact that about 50,000 gallons of oil and 41,000 pounds of Paris green dust mixture were utilized at a cost of approximately $25,000, the mosquito control was unsatisfactory, as is evidenced by an average count of 11 female *A. quadrimaculatus* per station per week and a maximum weekly count of 46 mosquitoes. The counts were not decreased greatly until larvicide measures were supplemented by marginal growth removal costing approximately as much as the larvicides. In 1936, irregular fluctuation was possible, resulting in some diminution in the expenditure for larvicides and a marked improvement in the station count (4 mosquitoes per station per week). In 1937, the combination of cyclical fluctuation and seasonal recession was employed for the first time. The decrease in larvicidal applications as well as in station counts (less than 2 mosquitoes per station per week), was dramatic. The story for 1938, 1939, and 1940 likewise confirms the importance of water level fluctuation as an anopheline mosquito control measure on impounded water.

Hinman's report includes a chart that shows graphically the relative costs and effectiveness of the antimosquito measures from 1935 through 1940 (see Table 4-2).

The benefits derived from the TVA program, as listed by Gartrell et al. (1972) include the following: malaria control; improvement in reservoir appearance by removal of debris; safer traffic for boats by removal of obstructions; benefits resulting from shoreline drainage for agriculture, public recreational areas, and upland game habitat; increased habitat for wildlife and migrating waterfowl in properly operated dewatering projects; deepening of shallow shoreline areas, permitting game fish better opportunity for feeding on forage fish; concentration of forage fish by drawdown so game fish can get to

TABLE 4-2 Cost and Effectiveness of Antimosquito Measures, TVA 1935-1940

Year	Water Management	Cost of Larviciding ($)	Number of *A. quadrimaculatus* station/week
1935	Constant pool level	29,000	11
1936	Fluctuation	23,000	4
1937	Fluctuation plus recession	5,000	1 1/2
1938	Fluctuation plus recession	3,000	2 1/4
1939	Fluctuation plus recession	3,000	2 1/4
1940	Fluctuation plus recession	5,000	2

SOURCE: Data from Hinman 1941.

them; shoreline clearance, which furnishes "openness" for Canada geese (the cleared areas can be planted with winter grasses, an important source of food for geese); piling cut brush above maximum pool level, affording excellent shelter for some species of upland game.

There are some unavoidable disadvantages: stumps and logs are natural resting places for wood ducks, turtles, shorebirds, and some amphibians and reptiles; marshes, which are attractive and productive, are destroyed by shoreline drainage; dewatering projects may reduce the production of food for waterfowl; deepening of shallow areas may destroy some fish spawning areas or areas favorable for production of food for fish and wildlife (however, these are only a minute portion of the total habitat available); water level management may interfere with spawning of some fish species (however, there is no evidence of significant problems); shoreline maintenance by cutting buttonball and willow adversely affects some species of wildlife, such as prothonotary warblers and yellow warblers, which use these for nesting.

The TVA (1973) *Environmental Statement* concludes: "No long-term environmental goals appear to be jeopardized by the vector control program on TVA reservoirs, and no significant cumulative impact on the environment can be identified."

The Crooked River Irrigation Project, Oregon

In the western United States, agriculture has become dependent upon increasingly extensive irrigation projects. These have created vast breeding habitats for dangerous vector mosquitoes and serious pest species of man and animals, notably *Culex tarsalis* and *Aedes nigromaculis*. In the Crooked River project, a task force was appointed by the Aquatic Plant and Insect Control Committee of the Pacific Northwest River Basins Commission, Water Resources Council (1968) to study the problem of mosquito propagation association with water resource projects in the Columbia Basin. Two seasons of field data were obtained prior to impoundment of the Prineville Reservoir, and additional information was available from the earlier studies by the Oregon State Board of Health. The multiple purposes of the irrigation scheme included irrigation, flood control, enhancement of fish and wildlife, and recreation. The reservoir, with a usable capacity of 155,000 acre-feet, is designed to supply water to 15,650 irrigable acres of the Ochoco Irrigation District and also to 7,660 acres of nondistrict lands. The system for delivering water to project lands consisted of diversion dams, pumping stations, pipelines, canals and ditches, and drains for land with imperfect drainage.

Prior to impoundment, sluggish marginal areas of the Crooked River were overgrown with algal mats and other vegetation, which were fertile breeding places for *Anopheles freeborni* and *Culex tarsalis*. Impoundment eliminated 79 percent of these areas. After impoundment, mosquito-producing areas in the reservoir itself were limited to approximately 80 acres of marshy, undrained depressions.

Within the irrigated areas, the mosquito breeding areas, prior to impoundment, consisted of natural or man-made surface pools, marshes, sloughs, and oxbows in the former channels of the Crooked River and its tributaries and miscellaneous waters such as seeps, drains, ditches, canals, and borrow pits. As a result of water management and installation of drains, 85 percent of the 2,433 acres of surface depressions, 49 percent of 796 acres of marsh, 80 percent of 583 acres of slough and oxbows, 79 percent of 380 acres of streambed pools, and 68 percent of 69 acres of miscellaneous waters were eliminated.

The report of the Crooked River project asserts that the benefits include a stable flow of water, elimination not only of heavy mosquito production but also of flood damage to agricultural lands, and creation of stream conditions favorable to trout. Assuming that the project has indeed brought about significant control, we can accept the estimates that the cost to control mosquitoes at $5.00 per acre by insecticides for a 50-year period without project development would be $1,065,250. The cost for 50 years after project development is estimated at $247,000.

One adverse consequence of the impoundment was the prolific production of blackflies in the stream channel for several miles immediately below the dam.

The Orleans Parish Mosquito Control Program

This program was started in 1964, with Mr. George T. Carmichael as administrator. The progress made in mosquito reduction is shown in the 1970 *Annual Report* of the Mayor's Advisory Committee on Mosquito Control:

Year	Average Number of Mosquitoes per Trap
1965	220
1966	105
1967	117
1968	78
1969	50
1970	37

Control activities were based on the results of larval and adult surveys, which revealed the location and extent of breeding areas, the times of adult emergence, and other information pertinent to the application of larvicides and adulticides. It was also recognized that, for many of the most prolific breeding areas, permanent control by source reduction would be advantageous. Each area is unique and requires its own specially designed system in which use is made of impoundments, weirs, dams, tidegates, drainage ditches, reservoir ditches, filling, and grading. By 1970, over 10,000 acres of once fertile breeding areas were under permanent control, and this number increased by 1971. "Small fill" areas were filled with waste material obtained without cost from the U.S. Gypsum plant. This material, in addition to leveling out depressions, so increases the alkalinity of the water that it becomes a deterrent to mosquitoes.

In the early part of the program, adults that escaped control measures in peripheral areas were controlled in urban areas by fogging. By 1971, with the exception of two units that were kept for special problems, the routine thermal fogging operations were phased out.

Water Resources and Mosquito Control in North Carolina

In North Carolina there are 206,350 acres of marshland, of which 47,500 acres are inland freshwater marsh; 100,450 acres are irregularly flooded saltwater marsh; and 58,400 acres are regularly flooded saltwater marsh. According to Mr. N. K. Oates (1970), Sanitary Engineer in Carteret County, the marshes regularly flooded twice daily by high tide provide nursery for plants, fish, shrimp, and other marine life, and they do not constitute a mosquito problem. The other type of salt marsh is flooded irregularly by rains and abnormally high tides, and in these mosquito breeding is exceedingly prolific.

Two types of salt marsh water management are used, ditching and impoundment. Enough ditches or canals are constructed so that the water is removed within 5 days after the marsh has been flooded. The ditches are about 5 feet wide and 4 feet deep, which allows water to remain in them all the time and so support marine life. Marine animals can move throughout the canal system. At the time of the report, some of the ditches were 10 years old and were holding up well. There had been no maintenance costs on the dragline ditches. It was claimed that this water level management brought about a 95 percent reduction of mosquitoes in the marsh. In Carteret County, there are about 70,000 acres of marshland, of which 15,000 acres had been brought under control.

The second type of control, impoundment, provides excellent mosquito control as well as habitats for waterfowl. Impoundments involve the construction of dikes, floodgates, and pumping stations. Soil from inside the impoundment is used for the dike; this creates deep water areas for marine life in winter and summer.

Mr. T. Stuart Critcher (1970), Assistant Chief of the North Carolina Wildlife Resources Commission, points out that marshlands are the most productive and fertile land areas in existence. They are essential to many kinds of wildlife and to shrimp, oyster, finfish, and crabs. By 1970, over 22 percent of the marshes had been altered by man, a significant amount of which was for mosquito control. Marsh alteration in four coastal counties is shown in the following table:

County	Total Acreage	% Altered
Carteret	48,600	10.2
Dare	16,000	43.8
Onslow	12,350	85.0
Pamlico	15,000	76.2
Total	91,500	36.9

Critcher says it is a matter of general knowledge that when a marsh becomes drier and plant succession advances to perennial woody species, the marsh becomes less suitable for muskrats, other furbearers, rails, and waterfowl and that ". . . the effects of ditching should be more thoroughly evaluated."

One advantage of impoundment referred to by Critcher is the increased opportunity for duck hunting. Balancing what the hunter is willing to pay for his recreation against the costs of impoundment, a benefit-to-cost ratio of 1.9 to 1.0 is estimated. If fishing is added, this becomes 3 to 1. "From a wildlife point of view, we can state that a salt marsh impoundment is one of the most highly productive types of wetland habitat . . . impoundments will all but eliminate the salt marsh mosquito . . . . But what are the other effects on the ecology of this type of environment?"

Elsewhere in this report it is noted that in all of the larger mosquito control districts in the southeastern United States, there is an increasing reliance upon salt marsh modification instead of larvicides, amounting to from one-fifth to one-third of the overall effort. This has been criticized severely by some environmentalists.

Florida: Relative Costs of Permanent and Temporary Mosquito Control on Salt Marshes

The cost of salt marsh impounding for mosquito control in Florida has increased from about $35 an acre in 1960 to $50 today. This does not include impoundments designed for multipurpose management to conserve all resource functions; one such impoundment recently constructed cost $168 per acre. Open ditch treatment of salt marshes, also done by dragline, commonly carries construction costs between $50 and $100 an acre. For comparison with larviciding costs, a figure of $60 an acre is therefore realistic for "permanent control" on the salt marsh. The cost of aerial larviciding varies considerably, depending on equipment and formulation. The intensity of treatment varies even more, as areas ditched or impounded had averaged from 3 or 4 to 15 larvicidal treatments a year. Annual cost of "temporary control" therefore ranges from $15 to $50 per acre, with $25 a fairly good average. It should be obvious then that in Florida the cost of impounding or ditching salt marsh is recovered in 2 or 3 years of savings on larviciding (see Table 4-3). The recovery is even faster if one also considers adulticiding costs, because under Florida conditions no larviciding of salt marshes ever completely obviates the need to adulticide.

A particularly careful analysis of control costs was made by the Indian River Mosquito Control District for a 175-acre mangrove marsh impounded in 1961. To the initial construction cost of $8,750 was added a dike-maintenance cost of $1,400 per year for a total 11-year cost of $24,150, or $12.54 per acre per year. The district's total 1972 expenditures for aerial larviciding, including a prorating of all general expenses, was $4.32 per acre of treatment. Since similar terrain near the impoundment requires an average of five treatments a year, the cost of larviciding the marsh over the same 11-year period would have been $41,580 or $21.60 per acre per year. Savings to the district over larviciding for 11 years were then $17,430 or $99.60 per acre. From this one can estimate that the 4,265 acres of salt marsh in the district impounded 10 to 15 years ago have already effected a saving of about $500,000 or nearly 2 years of present operating costs.

It was this sort of economy that led all of Florida's coastal mosquito control districts to impound or ditch all contained tidelands quickly once state aid was made available and restricted to such use by legislative action in 1953. A further inducement was the well-known fact that this source reduction totally abolished the salt-marsh mosquito nuisance, which even the best larviciding never accomplished.

TABLE 4-3 Relative Costs of Source Reduction and Larviciding on Florida Salt Marshes (Data from representative projects in six mosquito control districts)[a]

Type of Source Reduction, by County	Project Year	Project Acres	Construction Cost ($) Total	Construction Cost ($) Per Acre	Annual Larviciding Cost ($) Total	Annual Larviciding Cost ($) Per Acre	Year to Amortize Construction Cost by Saving on Laviciding
Impoundment							
Brevard	1973	102	17,145[b]	168	5,342	52	3.23
Broward	1961	231	9,021	39	7,357	32	1.22
Indian River	1962	175	8,750	50	3,780	22	2.27
Martin	1971	200	8,485	42	2,400	12	3.50
Ditch Systems							
Dade	1959	200	18,147	91	3,231	16	5.69
Ditches and Impoundments Combined							
Manatee	1970	220	11,250	51	7,383	34	1.50
TOTAL/ AVERAGE				441		168	2.63

[a]Data obtained from Florida Mosquito Abatement Districts.
[b]Cost was high because design was for multipurpose marsh management.

Source Reduction in California

According to various reports, California has some of the most serious mosquito control problems in the United States. After 25 years of insecticidal spraying, the pest and vector species have become resistant to all of the commonly used public health pesticides. The rich farmlands of California depend upon irrigation; with irrigation they produce 25 percent of the food eaten in the United States. This irrigation has also created vast and fertile mosquito breeding places in the agricultural valleys.

Mulhern (1973a,b) has divided the mosquito control activities into three phases. In the first, pre-DDT, phase the control operators utilized what tools were available and relied upon ditching as well as judicious use of larvicidal oils and Paris green. The second phase began with use of DDT, followed by the organochlorine, organophosphorus, and carbamate insecticides. The results were so beneficial that funds readily became available, and the programs were expanded to encompass more than 40,000 square miles, with a total expenditure of more than $10 million. The third phase, which Mulhern calls the "Era of Comprehensive Mosquito Control," is presently in its earliest stages. There will be restricted use of chemical pesticides in this phase and greater emphasis on source reduction and measures to induce landowners to prevent mosquito production on their properties.

R. H. Peters of the Bureau of Vector Control and Solid Waste Management said in 1971 that as a consequence of the reliance in recent years on pesticides, not one registered engineer was employed in California local mosquito control programs. He asserted further that although there are exceptions, "not one agency is at present going far enough . . ." in using methods of source reduction. Nevertheless, a number of schemes have been constructed for source reduction of specific problems. Peters (1971) tells of the modification of 30 miles of the Mokelumne River bottom by the use of caterpillar tractors, scrapers, ditches, and draglines. As a result of this program, which took 25 years for completion, it is now difficult to find *Aedes vexans* in this area of over 2,000 acres. Peters does not say what the effect was of destroying what he calls "natural jungles" on wildlife or native plant associations, but he admits that if the program had not started 25 years ago, there could have been objections because of possible adverse effects on the natural environment.

According to Robinson (1971) in Alameda County, the salt marshes have been almost eliminated, "the areas having been filled and now occupied by industry and warehouses. Land values in these areas are up to $15,000 or $20,000 per acre." No doubt many will agree that this is progress. In the Northern Salinas Valley Mosquito Abatement District, drainage has been employed since 1952. Greenfield (1971) asserts that the initial high cost of a drainage project is offset by savings in chemical control within 5 to 10 years. In his district, although mosquito problems have not been eradicated, they have been so reduced that they no longer are a costly control problem.

Irrigated pastures are an especially difficult problem. Lewis and Christenson (1971) describe a method of control by vertical drainage, in which holes were bored through the underlying hardpan to remove the excess water in depressions. Because open holes in pastures are hazardous to livestock, they were filled with "coarse aggregate" and topped off with a few inches of loam. The authors state that vertical drainage should result in greater tonnage of forage per acre. Additional observations in small pastures at Fresno State College permitted Lewis et al. (1972) to conclude that "vertical drainage holes drilled in problem areas and filled with a coarse aggregate can be a useful adjunct to mosquito control in areas where other water management procedures are impractical."

Junkert and Quintana (1972) in a discussion of the economics of pasture rehabilitation, point out that there are federal programs in which costs are shared with the landowners for conservation practices that protect the soil, reduce pollution, provide for more efficient use of waters and land, protect wildlife, retain open space, and beautify the landscape. Mosquito control is not included and so is not eligible for cost sharing. However, efforts are being made to include such control. Meanwhile, some cost-sharing projects designed for pollution control also reduce mosquito breeding. These include construction of terraces, ditches, and dikes; stream bank and shoreline protection; reorganization of irrigation systems; animal waste storage and diversion facilities; and tailwater recovery.

The current developments in California mosquito control are described as placing more and more reliance on source reduction, involving preventive measures in irrigation practices and enforcement of existing laws.

Conclusions

There are many engineering schemes that have been in operation for many years that have continued to control mosquitoes effectively in or near centers of population where such schemes are economically feasible. It would be desirable to obtain more detailed information on the costs of installation and maintenance, the extent to which they have brought about disease and pest control, and the costs of obtaining comparable control by the use of chemical insecticides. There should also be a comparison of the side effects of both types of control, including destruction of natural environments and wildlife, esthetic values, and agricultural gains.

W. B. Herms is quoted by Peters (1971) as follows: "If you can get rid of the water that produces mosquitoes, you won't have mosquitoes." The question is whether one wishes to get rid of the water.

Each mosquito problem is unique and will require evaluation by the combined efforts of persons trained in epidemiology, entomology, engineering ecology, and economics. Such a group would be best qualified to determine whether mosquito control should be undertaken at all, and if so, which method or combination of methods would be the most economical, the most beneficial, and the least destructive to the environment and to its plant, animal, and human inhabitants.

BIOLOGICAL CONTROL

Biological control is "the action of parasites, predators and pathogens in maintaining another organism's density at a lower average than would occur in their absence" (De Bach 1964). Biological control is a part of the natural controls that limit the numbers of all organisms. In the context of pest control, biological control is a natural ecological phenomenon that may be utilized in attempts to maintain pest populations at satisfactory levels. Some scientists and others have a much broader definition of biological control that embraces use of host resistance, competitive displacement, autosterilization, genetic manipulation, habitat management, and biologically produced chemical compounds. However, for the purposes of this report, the narrower traditional concept will be used.

In recent years, there has been great interest in expanding the utilization of biological control in the control of all pests, especially insects (NAS 1973). For this reason, the study team gave considerable attention to the role of biological control in the management of populations of mosquitoes and other arthropods of health importance. Parasites, predators (both vertebrate and invertebrate), and pathogens take a tremendous toll on mosquito populations. If it were not for the impact of their natural enemies, the problems of controlling these important pests would be even greater. In many instances, the control resulting from the natural enemies is not sufficient to meet the criteria set by man (e.g., acceptable population levels of a pest mosquito), and other control measures must be instituted. However, in most situations current knowledge is inadequate to assess the value of biological control or to devise ways in which its partial control can be best utilized (NAS 1973).

There are almost no cases where development of resistance has interfered significantly with successful biological control, and the biological control agents have no unwanted side effects such as toxicity or environmental pollution. Where applicable and effective, the utilization of natural enemies is an ideal pest control technique.

Setting aside a few notable exceptions, e.g., *Aedes aegypti* and *Anopheles gambiae*, the insects of medical importance are indigenous to the areas they currently occupy. Hence, the approach of classical biological control is largely excluded for these insects. Biological control for these pests involves primarily the manipulation and fostering of natural enemies already present in the environment. An exception may exist for general predators or other natural enemies that are catholic in their host preferences. However, even in these cases, the introduced natural enemy will most often need to compete in niches already filled, and successful and effective establishment will be difficult. It is for these reasons that biological control of insects of medical importance necessarily has emphasized the utilization of natural enemies that can be mass-produced and then introduced into critical places in the environment or at critical times of the year. This approach, of course, requires the continued maintenance of mass-rearing facilities and an adequate distribution system for the natural enemy.

The evaluation of costs of biological control depends very much on the permanence of the control established. If the introduction of a new natural enemy provides control (or partial control) indefinitely, then the cost of the control (even if the initial cost of control is high) will be extremely low. However, not all biological controls have such permanence, and they sometimes have to be re-established, e.g., restocking a temporary pond with *Gambusia*, or re-establishing a parasite after an unusually severe winter.

A number of factors affect the success or failure of the natural enemies, and these in some situations limit their usefulness. In some situations, current knowledge does not provide any manageable and useful enemies. However, where natural enemies can be utilized, they are usually of low cost, make significant contributions to control, and rarely have undesirable side effects. The potential and hazard of insect parasites and predators, fish, pathogens, and nematodes as biological controls for arthropods of health importance are discussed in more detail in the following sections. A number of reviews that treat this topic are available in the scientific literature, e.g., Jenkins 1964, Laird 1970, 1971, Brown 1972, and Chapman 1974.

In investigations extending over more than a half a century, a wide variety of biological controls have been studied. A few of these--e.g., larvivorous fish--have become an effective and established control procedure where the mosquitoes exist in limited environments. A large number of other biological control agents have been identified and in a few instances their characteristics and limitations described, but the full potential of these agents is largely unknown (NAS 1973, Smith 1973).

Insectan Parasites and Predators

The insectan parasites and predators have not been as important in the control of mosquitoes and other arthropods of relevance to health as they have been with agricultural pests. A wide array of insectan predators, including damselflies, water boatmen, back swimmers, giant water bugs, predaceous water beetles, ants, and mosquitoes, attack mosquitoes and other arthropods (Bay 1974, Jenkins 1964, Laird 1971). A number of other invertebrate predators (e.g., *Hydra*) contribute to this heavy natural mortality of mosquitoes.

However, very few if any of these predators are manipulable in large vector control programs. The genus *Toxorhynchites*, a group of large nonbiting mosquitoes predaceous as larvae, and the notonectids, or back swimmers, appear promising and are under investigation as possible predators that might be manipulated in a management program, but these must be still considered only as experimental efforts at this time. At best, they will probably have a place only in limited environments. *Toxorhynchites* would appear to be especially promising against mosquito pest populations that breed in small containers, often in obscure, difficult-to-find places--discarded tin cans, old tires, flower vases, tree holes, and the like--where the application of an area spray with insecticide would be extremely wasteful. The released *Toxorhynchites* females seek out such places to lay their eggs. In nature, the rate of development of these predators is slower than that of the pest mosquitoes; hence, to offer effective biological control, they must be mass-reared and released each year at a critical point in the season (NAS 1973, Corbet and Griffiths 1963, Trpis 1973). Additional field trials with these predaceous mosquitoes are under way.

The use of insectan parasites or parasitoids would seem to be even less promising at this time. A few telenomid egg-parasites that attack the important disease vectors of the genus *Triatoma* have been discovered and are currently under investigation, but they do not provide a control at this time. Mosquitoes appear to lack insectan parasites completely.

Fish

It is clear from the literature (Bay 1967, 1973; Gerberich and Laird 1968) that the biological control of mosquitoes by fish means primarily the purposeful introduction of larvivorous fish into mosquito-breeding waters and only peripherally the management of endemic larvivores. The practice is old and stems from repeated observations, in field and laboratory, that certain small fish are voracious feeders on mosquito larvae and pupae. Although it is estimated that over 1,000 species of freshwater fish consume mosquito larvae (Myers 1973:65), attention very early (about 1904) fastened on the American live-bearing minnow, *Gambusia affinis*. In the 1920s and 1930s this "mosquito fish," as it became known, was introduced for mosquito control not only in parts of the United States where it was not native but throughout the world. It was joined by a South American live-bearer, the common guppy, *Poecilia reticulata*. Although many other fish, especially in the families Cyprinodontidae (killifishes) and Poeciliidae (live-bearers), have been added to the biological control armamentarium, gambusia and guppy remain the chief larvivores introduced for mosquito control (Wright et al. 1972).

The merits of *Gambusia affinis* were first recognized in malaria control campaigns in the southeastern United States where it was native and common. As an inhabitant of permanent waters and a surface feeder, it was particularly effective against anopheline mosquito larvae. Its reputation spread fast, and its remarkably rapid dissemination throughout the world was largely as an adjunct to other malaria control procedures. It was early realized, however, that its ability to live and propagate in the confined waters of artificial containers, from urns and cisterns to shallow wells and garden pools, made it an excellent control agent against *Aedes aegypti* also, and it was used extensively in the yellow fever campaigns during the building of the Panama Canal and in the eradication of this disease from Havana at the turn of the century

(Gerberich and Laird 1968). It is significant that among the world's hundreds of larvivorous fish species, the bias in favor of *Gambusia affinis* is attributable to its success in malaria and yellow fever control. The characteristic tolerance of pollution in *Poecilia reticulata* made it in Asia an analogous tool against filariasis, whose chief vector, *Culex pipiens fatigans*, has a predilection for polluted waters.

Against culicine mosquitoes, gambusia is not a markedly superior predator (Harrington and Harrington 1961). It is low on cold tolerance, and it is readily isolated from its prey by algal mats and aquatic weed growths, obstructions easily overcome by its companion and equally voracious larvivore, *Heterandria formosa*, a very small live-bearer also once called the mosquito fish (Innes 1917, Myers 1973:65).

It is frequently recommended that obstructive plants be removed to facilitate the work of gambusia, but in many parts of the world, particularly in India and southeastern Asia, such vegetative clearing also facilitates predation by a host of indigenous species (e.g., *Panchax* spp.), so that exotics like gambusia are superfluous. Weed removal by co-introducing *Tilapia* species, carp, *Cyprinus carpio*, or white amur, *Ctenopharyngodon idella*, is also practiced, but all these plant-eating fish are considered risky introductions for ecological reasons. Gambusia, too, is an ecologically dangerous introduction into natural habitats (Myers 1965), and its future use as a biological control agent will have to be more circumspect than in the past. It will nevertheless continue to be used, as witness its current popularity in California rice-field mosquito control (Hoy and Petersen 1973) and the excellent guide for its use prepared by the U.S. Navy (Sholdt et al. 1972).

For mosquito control in temporary waters, it has been proposed (Hildemann and Walford 1963) to use annual or "instant" fish, such as *Cynolebias* and *Nothobranchius* species, which bury their eggs in the bottom of drying waters, the eggs then surviving considerable drought and hatching immediately on flooding. Several *Cynolebias* species are currently under investigation for mosquito control in the rice fields of California (Hiscox et al. 1974). Results are promising, and there are indications that environmental impact from escapes to natural waters would be minimal.

As a control technique, the stocking of larvivorous fish is often less costly than repetitive larviciding with oil or insecticides. It is especially inexpensive if the fish are already established and thriving in some nearby waters and need only to be collected and transported to where they are needed. Ingenious local methods are usually developed that make the operation simple. Some mosquito control districts combine helicopter inspection with control by simply dropping plastic bags full of fish into the scattered marsh pools that are found to be breeding mosquitoes. However done, successful stocking can reduce both larviciding and inspection for a long time, thus effecting a substantial saving in cost. Where the technique is used on a large scale, it is sometimes found economical for the program to rear larvivorous fish for dissemination in its own fish ponds, custom-designed for the purpose. Several control districts in California found this profitable even at the height of DDT's effectiveness and inexpensiveness, and today five of that state's districts are actively engaged in gambusia-rearing investigations (Hoy and Petersen 1973).

Control by stocking of larvivores will probably grow in practice as chemical larvicides are de-emphasized. It is often recommended that the two techniques be used together, the one reinforcing the other. In any event, it is clear that chemical larviciding should not injure larvivorous fish, whether introduced or native. Most larviciding methods in use today or being developed are tested for possible hazard to larvivorous fish. This will continue, probably on a larger scale, in the future. Another threat to the use of larvivorous fish is posed by predators on them. These can be stocked game fish, as has been observed in Texas, or escaped exotics such as the rapacious *Belonesox belizanus*, a large poeciliid now a problem in south Florida because it is a prolific live-bearer whose diet includes gambusia and other top-minnows.

The natural regulation of mosquito numbers has never been studied in sufficient depth to permit quantitative appraisals of predator, parasite, or pathogen roles, yet it is commonly assumed that predatory fishes are extremely important. This assumption is supported by the effectiveness of "minnow-access" ditches on America's salt marshes, thousands of miles of which have been constructed and maintained over the past 60 years. The success of such ditching in freshwater as well as saltwater areas is documented in hundreds of published reports not mentioned in Gerberich and Laird's (1968) bibliography, most likely because the augmentation of endemic, natural controls is not traditionally considered biological control. Endemic larvivorous fish nevertheless constitute a largely unexploited source of help in mosquito control, awaiting the necessary research and development.

In summary, problems in the biological control of mosquitoes by the use of fish fall into two categories. First, in man-made breeding waters, the introduction of exotic larvivorous fish will continue to be useful and will demand more exploration and investigation to discover ideal species for specific situations and climates. It will be constrained by environmental considerations, such as the possible injury to native biota and ecosystems by escape of exotics into natural waters. Thus, more thorough studies of potential environmental impacts will be required in advance of new introductions. Second, in natural waters, indigenous larvivores have scarcely been exploited as control allies except on the salt marsh. The research that will open up this alternative to drainage and larvicides in freshwater habitats is mostly in the future. Only the study of population dynamics in sufficient depth will uncover or suggest habitat manipulations that can augment the regulatory role of indigenous larvivores.

Pathogens

A variety of pathogens, including fungi, nematodes, protozoa, bacteria and viruses, offer great promise in the control of some arthropods of medical importance (Laird 1970, NAS 1973). However, before these can come into general use, it must be determined that they are safe for humans and domestic animals (WHO 1973, Smith 1973).

Bacteria

Various strains of the crystalliferous spore-forming bacterium *Bacillus thuringiensis* have been commercially available for some time for use in the control of lepidopterous pests of agriculture and forestry. The pathogen acts both through action of its crystalline toxin and the further multiplication of the bacilli in the insect body (Heimpel 1967). Unfortunately, these commercially available strains have little or no pathogenicity for mosquitoes. A variant (bacillus BA-068) of *Bacillus thuringiensis* that has been isolated from *Culex tarsalis* is especially effective against mosquito larvae (Reeves and Garcia 1971), but this strain is not yet commercially available. *Bacillus thuringiensis* may also have a limited role in housefly control. This bacterium is generally considered to be environmentally acceptable, and there is no indication of harmful effects on man or plants (WHO 1973, NAS 1973). Other bacteria, such as *Pseudomonas*, may produce toxins and make breeding sites free of mosquitoes but these do not appear to have much promise at present. Much additional research and development will be necessary before these can be brought to the point of field trials.

Fungi

More than 500 species of fungi have been reported to attack insects, and some have been used effectively in pest control. The entomogenous fungi are extremely diverse, and generalizations about them are difficult to make. Although some fungi have a commensal or mutualistic relationship with insects, the ones of concern here attack the insect tissue and cause death; in some cases death may result from a toxin produced by the fungus (Madelin 1966). *Beauveria bassiana* is mass-produced for the control of leaf-feeding insects and *B. tenella* and *Metarrhizium anisopliae* appear promising for the control of subterranean beetles. *Coelomomyces* spp., *Lagenidium giganteum*, and *M. anisopliae* appear to be very promising for control of mosquitoes in some situations and are being studied intensively. The attack of *Beauveria* and *Entomophthora* on houseflies is well known but with present knowledge does not provide significant control.

Coelomomyces spp. are aquatic fungi that are obligate parasites in mosquitoes and a few other insects (Couch and Umphlett 1963, Couch 1972). *Coelomomyces stegomyiae* has been used successfully in at least one field trial (Laird 1967, 1970). Although promising at this time, the use of fungi in mosquito control is still in the experimental stage (NAS 1973). Much work remains to be done, especially on mass production, effectiveness under field conditions, and impact on nontarget organisms (WHO 1973).

Nematodes

Some species of mermithid larvae (roundworms) enter mosquito larvae and kill them (Welch 1962). The most promising of these, *Reesimermis nielseni*, has been cultured in large numbers, and preliminary field trials gave infection rates ranging from 50 to 85 percent (Chapman 1974). This nematode attacks over 50 species of mosquitoes, especially *Culex pipiens fatigans* and *C. tritaeniorhynchus*. It is not anticipated that the use of these host-specific nematodes would create any hazard.

Protozoa

Among the protozoa, the microsporidia, especially several in the genera *Nosema* and *Thelophania*, appear to be the most important in causing disease in mosquitoes (Chapman et al. 1970). The status of host specificity in this group must be further determined before these organisms can be used in pest control. Nevertheless, they offer much hope for the future. The possibilities of using *Nosema*, *Thelohania* and *Pleistophora* against various mosquitoes under field conditions is being studied in several places (Reynolds 1972, Chapman 1974), but they all must be considered experimental. Much research on toxicity to mammals and other nontarget organisms, mass production, and effectiveness must be completed before they can become practical controls (NAS 1973, WHO 1973).

Viruses

The virus pathogens, especially the nuclear polyhedrosis viruses (NPV) and granulosis viruses (GV), are among the most promising of microbial control agents for agricultural insect pests, especially when considered from viewpoints of specificity, efficacy, and safety. Recently, several viral agents attacking mosquitoes have been identified (Chapman 1974). The first confirmed NPV (nuclear polyhedrosis virus) from mosquitoes was isolated from *Aedes sollicitans* (Clark et al. 1969). This virus also attacks other mosquitoes of the genera *Aedes*, *Psorophora*, and *Culex*. It is a very virulent virus and offers much promise. Other mosquito viruses, including cytoplasmic polyhedrosis viruses, irridescent viruses and a poxlike virus, are known. However, because of possible problems of toxicity to nontarget organisms and development of mass production, it is believed that it will be several years before these viruses can go to large-scale field trials (NAS 1973, WHO 1973).

One of the critical stumbling blocks in the development of insect pathogens for use in pest control has been their clearance with respect to safety and possible hazards. This topic was discussed in considerable detail by a group of virologists, toxicologists, ecologists, and insect pathologists convened by FAO and WHO in Geneva, Switzerland, in November, 1972 (FAO/WHO 1973). They concluded, among other things, that the nuclear polyhedrosis viruses and the granulosis viruses offer the greatest potential as control agents in terms of efficacy and safety.

Of special significance with respect to consideration of a special case for NPV and GV are these three points: (a) there is no confirmed evidence that these viruses can multiply in vertebrate cell cultures; (b) their range of specificity is narrow (the NPV and GV are only associated with Class Insecta; specific viruses are usually restricted to a single genus or a group of genera); and (c) their morphology is unique among the viruses (they are contained within a proteinaceous matrix, i.e., in the case of NPV, the polyhedron or inclusion body).

A similar conference sponsored by WHO in April 1973 reviewed the safety of other biological agents for arthropod control (WHO 1973, Smith 1973). The safety of use varies widely from pathogen to pathogen, depending on the nature of the use, the characteristics of the pathogen, and limitations in our knowledge of these pathogens.

GENETIC CONTROL

This technique, based primarily on the release of radiosterilized or chemosterilized males of the species, has attained success in the control of the screwworm fly, an insect of veterinary importance. Among insects of public health importance, tsetse flies have been the subject of much work preliminary to genetic-control operations, and houseflies have been almost completely eradicated with chemosterilant baits in certain locations in Florida and a small Caribbean island. The only vector species that have been subject to extensive field trials are the mosquitoes (Pal and LaChance 1974).

The first attempts made with radiosterilized males against *Anopheles quadrimaculatus* and *Aedes aegypti* in Florida were unsuccessful, due to insufficient numbers, lack of mating competitiveness in the liberated strain, and reinfestation. The use of chemosterilized males did attain eradication of *Culex quinquefasciatus* on Seahorse Key off the Florida coast (Patterson et al. 1970), although specimens of this mosquito were accidentially reintroduced by boats coming to the island. The use of males of a strain cytoplasmically incompatible with the indigenous population successfully eradicated *Culex fatigans* (*quinquefasciatus*) from the isolated village of Okpo in Burma (Laven 1967). The liberation of males and females of a strain carrying chromosomal translocations has eliminated a small population of *Culex pipiens* at Monpellier, France (Laven et al. 1971). A large-scale program carried out by the WHO/ICMR Unit on Genetic Control of Mosquitoes at New Delhi, India (Smith 1972), has already compared the three sterile-male methods (radiosterility, chemosterility, and cytoplasmic incompatibility) for genetic control of *C. fatigans* (Rajagopalan et al. 1973) and at present is placing the main emphasis on males treated with the pupal chemosterilant thiotepa (Sharma et al. 1973).

The sterile males obtained by hybridizing two species in the *Anopheles gambiae* complex, *Melas* X (Species B), were released in a village in Upper Volta infested by *A. gambiae* (Species A); the results were sufficiently promising to warrant consideration of a repeat experiment with a hybrid containing Species A (Davidson et al. 1970). Excellent results have been obtained with chemosterilized males of *A. albimanus* released over a 5-mi^2 area in El Salvador.

Essential for success in the technique is sufficient production of the sterile males. For a sizable project directed against mosquitoes, the production should be of the order of 5 million a week, to maintain the liberated number around 10 times that of the males of the target population. The necessary fieldwork to study the numbers, behavior, and ecology of the target population and to perform and monitor the releases requires high scientific competence. The greater part of the successful work in genetic control of mosquitoes derives from a single laboratory, that of the USDA at Gainesville, Florida, and it has been their training and influence that has enabled the projects to be performed at San Salvador and New Delhi, where national governments have made adequate laboratories and other facilities available.

Genetic control is regarded by some as the most promising of the alternatives to chemical control of vectors, but its full potentialities and hazards have yet to be assessed (Whitten and Pal 1974). The use of translocation strains, which are particularly suitable for *Aedes aegypti* control (Rai et al. 1973), is especially in need of evaluation, as there is a danger of their perpetuating the transmission of the arbovirus that is the ultimate target. The logistical difficulties involved in attempting control over vast mainland areas are formidable, as the pressure of reinfestation is omnipresent. It is nonetheless possible that genetic control of vectors will develop into a flourishing activity, steadily gaining in the expertise resulting from its exploitation in agriculture as well as public health. But the place of genetic control will remain in the governmental rather than the private sector, at least until it enters the realm of corporation contracts.

COMPETITIVE DISPLACEMENT FOR MOSQUITO CONTROL

Competition has been an important factor in determining the relative abundance and distribution of animals. De Bach (1966) states that "different species which coexist indefinitely in the same habitat must have different ecological niches, that is, they must not be ecological homologues." If one of the species can utilize marginal habitats, nutriment, or other basic needs, an equilibrium between the two populations may be established. There has been much controversy regarding the mechanisms of competitive interaction, and some authors consider that predation, fighting, and other physical harassment is an essential part of the interaction. Theoretically, if the two species occupy exactly the same niche, and this niche can support only a finite number of individuals, then one of the two competitors should become the sole occupant simply by producing more surviving offspring.

There have been many laboratory demonstrations of competitive displacement of one species by another, including the classic experiments of Park and his coworkers with grain beetles. Gubler (1970) observed that in small cages in which *Aedes albopictus* and *A. polynesiensis* were competing for limited space and food, the latter species disappeared. Lowrie (1973) demonstrated that when competive pressures were exerted only during the larval stages, *A. albopictus* again eliminated *A. polynesiensis*.

According to De Bach, there are many cases of actual or apparent competitive displacement in nature, but the only one that has been thoroughly documented involved several species of *Aphytis*, ectoparasites of the California red scale. There have been limited observations on this phenomenon involving mosquitoes in nature. *Anopheles atroparvus* in northern Europe breeds in salt water; *A. messeae* occurs in fresh water. Hackett (1937) said, "*Atroparvus* is not able to compete with *messeae* in fresh water in Holland." This, indeed, is fortunate, for *A. atroparvus* was the important malaria vector. The question is whether the larvae of this species are not able to live in fresh water and for this reason are not found with *A. messeae*. The fact that in Spain and Portugal *A. atroparvus* breeds in fresh water, and also that it can be reared in the laboratory in fresh water, indicate that there is more than the salinity factor that excludes it from fresh water in Holland.

Toxorhynchites brevipalpis and *T. amboinensis* were introduced into Hawaii in 1950 and in 1968, respectively.

Steffan (1970) reports that in recent years he has been able to find only *T. amboinensis*. Because both occupy the same habitat (tree holes and similar small water containers), Steffan suggests that *T. brevipalpus* was eliminated through competition with *T. amboinensis*. If so, no doubt predation was a major factor.

Covell and Resh (1971) suggest that the reversal in the relative densities of *Culex pipiens* and *C. restuans* larvae in catch basins in Jefferson County, Kentucky, may in part be caused by competition.

It is possible that the invasion of Southeast Asia by *Aedes aegypti* has had an effect on the abundance of the indigenous *A. albopictus* in urban areas. *A. albopictus* breeds in small natural water containers, but especially where *A. aegypti* is absent (e.g., Honolulu, Guam) it is abundant in artificial containers in close proximity to houses. When *A. aegypti* also is present, there appears to be an inverse relationship between the density of *A. albopictus* and proximity to houses. The opposite is true of *A. aegypti*.

In bamboo thickets and coconut groves in the Philippines, *A. albopictus* is predominant, although in some places small populations of *A. scutellaris* may be found. In Melanesia and Polynesia, *A. scutellaris* and its relatives exist in dense populations in coconut groves; here there are no *A. albopictus*.

These indications that competitive displacement may limit certain mosquito populations suggest the possibility that under proper conditions a vector species may be replaced by a nonvector. Such conditions would include precautions that the introduced species is not in itself capable of introducing a new disease into the target area or of exacerbating the pest problem.

NEW MATERIALS WITH BIOLOGICAL ACTIVITY

It should be stressed that the new materials of novel action, such as the insect growth regulators and overcrowding factors, being organic chemicals just as the conventional insecticides are, are just as liable to select out strains with biochemical defense mechanisms making for resistance. The development of populations that no longer respond to attractant chemicals employed against them would appear unlikely, although not impossible.

Insect Growth Regulators

This term is applied to a number of new compounds that do not kill directly, but inhibit the development of an insect at some point before it becomes a mature adult; the term "insect development inhibitors" is also used and is more appropriate. The principal group of compounds under this heading are the juvenile hormone (JH) mimics. Mosquito larval control is among the most promising uses for these compounds (Sacher 1971, Speilman 1972), and one (Altosid, common name methoprene) has now been registered for this use. Because the JH mimic is effective against the larvae only when they are at a stage when their "window" is open for its action, namely the last larval instar, emphasis is placed on slow-release formulations (Dunn and Strong 1973).

Although insecticide-resistant strains of several pest insects show some cross-resistance to JH mimics, it is insufficient to be practically important. Methoprene is completely effective at an active-ingredient dose of 0.125 lb/acre to control multiresistant *Aedes nigromaculis* in California (Schaefer and Wilder 1972) and, if applied to even-aged populations shortly before they pupate, 0.02 lb/acre is completely effective. Although selection pressure from methoprene for 20 generations has induced no resistance in *Culex quinquefasciatus* larvae (Schaefer and Wilder 1973), pressure on a *C. pipiens* strain has produced appreciable resistance in only three generations of selection.

An insect growth regulator that has achieved very successful control of mosquito larvae is the fluorinated urea derivative PH 6040 or TH 6040 (Largon). Its mode of action involves the mobilization of chitinase to prevent the proper formation of the larval, pupal, and adult cuticle. The cross-tolerance of organochlorine- and organophosphate-resistant strains to Largon is negligible (Jakob 1973).

A considerable number of aliphatic amines are possibilities as larvicides (Mulla 1967, Mulla et al. 1970, Cline and Hall 1973), and four have been registered for this use in California. Their mode of action is apparently due to their disruptive action on the water balance and cuticular functions of the aquatic stages; they have a delayed effect as well as the initial direct one. They are effective against eggs as well as pupae, and a mixture of 1.0 percent ethanolamine with 0.1 percent dodecylamine has been found to be an effective ovicide for *Aedes aegypti* (Cline et al. 1969) (there is special need to destroy the diapausing eggs of this vector, which can survive a long period of desiccation in used tires, the most common source of reinfestation of an area) (Jakob et al. 1970).

Attractants and Pheromones

Among insects of public health importance, the housefly is the species most responsive to attractants. Although a wide range of chemical compounds are known to attract them, simple materials such as molasses and grain alcohol have been used in traps; the bodies of the flies already trapped produce attractants to draw in more flies. In practice, attractants are used in baits containing organophosphorus insecticides such as trichlorfon and dimethoate, molasses and corn meal being favored as inexpensive and easily obtainable. Attractants for adult tabanids and simuliids are usually visual, in the form of traps based on response to light-and-shade; for simuliids, CO_2 increases the trapping power, and a specific chemical attractant is known for *Simulium euryadminiculum*. Light traps have been widely used for mosquitoes, but for surveillance rather than control. They are most effective for *Culex*, which is particularly responsive to the addition of CO_2. Of specific attractants, the only significant compounds for adult mosquitoes are lactic acid (Smith et al. 1970), bromopropionic acid (Carlson et al. 1973), and lysine (Brown and Carmichael 1961). The search for natural pheromones of mosquitoes has yielded evidence of a chemical stimulant for sexual behavior in *Culiseta* (Kliewer et al. 1966) and a contact pheromone enabling males to recognize females of the same species in *Stegomyia* (Nijhout and Craig 1971).

Overcrowding Factors

Overcrowding factors were discovered to be produced by *Culex fatigans* larvae when crowded to the extent of at least 500 larvae to 100 ml of water, and they have the effect of killing newly hatched larvae (Ikeshoji and Mulla 1970a). Depending on their concentration, they take between 1 and 6 days to kill, mortality usually occurring at ecdysis. Chromatography of an ether extract of the water yielded two active zones (Ikeshoji and Mulla 1970b); four active compounds have been chemically identified.

REPELLENTS

An important measure in the prevention of arthropod-borne infections is the protection of individuals or even large groups of people whose activities require their presence in areas where they are exposed to the vector species. Also, in areas with dense pest populations of mosquitoes or biting flies, the use of protective clothing and repellents may be highly desirable if not essential. Especially during World War II, there was an intensive screening program for compounds that would be more effective than the pine oil and citronella that were used in earlier times. The three most promising were found to be dimethyl phthalate *n*-butyl mesityl oxide oxalate (Indalone), and ethyl hexanediol (Rutgers 612). Other effective substances were benzyl benzoate, dibutyl phthalate, dimethyl carbate, and butyl ethyl propanediol. A synergistic effect was obtained against a broader spectrum of insects by combining several of these materials. The liquid repellents are smeared onto the exposed skin or rubbed into the clothing. Clothing may be impregnated by being dipped into or sprayed with properly formulated emulsions or solutions.

In the continued search by the USDA for more and better repellents, it was found that N,N-diethyl-*m*-toluamide (deet) has many advantages over other compounds (Gilbert et al. 1957, McCabe et al. 1954). It is more effective against a wider range of bloodsucking arthropods; it can be diluted with alcohol and so is cosmetically more acceptable and is less damaging to fabrics and plastics. It is available commercially in both liquid and aerosol formulations.

The above compounds are contact repellents--that is, the arthropod is repelled when it lands on the surface of the skin or clothing that has been treated. There is also a search for space repellents that prevent the insect from alighting because of a vapor. Gouck et al. (1971) treated wide-mesh netting with a large number of compounds and tested them against salt-marsh mosquitoes. A number of these afforded 90 percent protection for about 3 months. Fine-mesh bed-nets or head-nets may become most uncomfortable, especially in the tropics. Wide mesh netting, allowing a greater circulation of air, should be a great improvement if the vapor repellents afford significant protection against flying insects. Such netting can be used in open windows or as tent-like cover for people sleeping outdoors.

No satisfactory systemic pest repellents have been found as yet.

NEW WAYS OF USING CONVENTIONAL PESTICIDES

The new insecticide practices being developed are not radically novel, but are based on existing practices. ULV methods of spraying are now in widespread use. This method exploits the fact that organophosphorus insecticides either are themselves liquids or can be obtained in concentrated solution that may be applied from fixed-wing aircraft (Kilpatrick et al. 1970) or helicopters (Hayden et al. 1973). The insecticide penetrates into the sheltered resting-places of adult mosquitoes because it is atomized into very small droplets: about 25 microns in diameter is the most efficient size (Weidhaas et al. 1970). If no large droplets are accidentally produced, there will be no spotting of the finish of automobiles. A considerable number of ground machines have been developed for ULV application (Mount el al. 1972), and fogging machines may be adapted for this purpose (Anderson and Schulte 1970). The mist produced is less visible than thermally generated fogs, and gives a control that lasts longer after application. Gram for gram, it is more effective than conventional sprays and does not wet the clothing of people who might be in the area during the treatment. However, under some circumstances ULV has the disadvantage of excessive drift.

There have been developments in the ways in which insecticides may be applied in solid or granular form, to give a slow and continuous release of the active toxicant. Granules on clays or bentonite were originally developed for mosquito control in reedy swamps. The revival of Paris green for culicine control involves mixing it with vermiculite and an emulsifiable oil (Rogers and Rathburn 1960). There has been a recent stress on sand granules, which leave the treated water entirely clear, e.g., treatment of drinking-water jars for control of *Aedes aegypti*. Slow-release formulations are particularly important for larvicides of the juvenile-hormone mimic type (e.g., Altosid, a commercial product of methoprene) to ensure that there is always active compound present at the time shortly before the larvae are due to pupate. The U.S. Army Environmental Hygiene Agency has obtained excellent results with 2-6-mm^3 pellets of organophosphate larvicides such as chlorpyrifos in plastic polymers (Nelson et al. 1973) and with encapsulated formulations of malathion or Abate in these polymers (Miller et al. 1973); these matrices, however, are not biodegradable. Good results have been obtained against larvae of three species of *Culex* with tributyltin oxide impregnated into rubber cubes (Boike and

Rathburn 1973), and rubber sheets are being investigated for control of *Simulium* in running streams (N. A. Cardaielli University of Akron, Personal communication).

A significant change in methods of mosquito control has been the integration of chemical larvicides with the use of larvivorous fish such as *Gambusia* and *Poecilia*. Quite widely developed in California, Utah, and other states, the best example is on Oahu island, Hawaii, where the successful control of the culicine mosquitoes produced by the large Kawainui swamp is assured by applying at the end of each rainy season a single treatment of fenthion to cut the larval densities down to a size that can be handled by the population of several fish species maintained there (Nagakawa and Ikeda 1969). In the Delhi area, India, control of *Culex fatigans* breeding in wells is attained by *Gambusia* integrated with applications of Abate.

QUARANTINE PROCEDURES

Quarantine procedures directed against the arthropod vectors of human disease have always been difficult to apply. The plant quarantine system that was developed in the late decades of the nineteenth century and the early part of this century has had little influence on the spread of arthropods of public health importance. With the tremendous increase in air traffic in recent times, it has become even more difficult to limit the spread of vectors. Fortunately, most of the arthropod vectors are so closely adjusted ecologically and behaviorally to their habitats that their ability to spread to new areas and become successfully established is greatly limited (there are exceptions to this generalization--e.g., *Aedes aegypti* and *Anopheles gambiae*). This has been our major source of protection, because the quarantine procedures themselves have little impact on arthropod vectors.

One quarantine aspect that has been neglected in vector control is control over the movements of larvivorous fish. These introductions have been deliberate and often with little consideration of impacts of these fish on the indigenous fish. Introduction of fish into new areas should be made with appropriate evaluation of the impact of these introductions on the endemic biota and the possible hazard of contamination of the fish introduction with other, undesirable species.

ERADICATION VERSUS CONTAINMENT

The tools used to control disease vectors, including conventional pesticides, as well as the various alternatives discussed in the preceding sections, must be put to work within the context of an overall pest control strategy. Two approaches are possible once a pest has become established in a particular locality: eradication, where an attempt is made to eliminate the pest completely; and containment, where the pest population is kept at an acceptably low level by an ongoing program of control.

The goal of eradication of an infectious disease is the complete cessation of transmission: there can be no more new infections. Eradication of an animal species means that the last individual has been exterminated. These criteria ultimately must have a global application, for as long as one case of an infection, or one breeding pair of an animal species survives anywhere in the world, they may eventually be the source of a recrudescence of the disease in a nonimmune population, or of a reinvasion of clean areas by the animal pest. The eradication of a disease on a global basis is extremely difficult. However, the application of the principle of eradication to more limited objectives may achieve significant results. The vector or pest may be eliminated from a more or less limited geographical area that can be protected from reinvasion. This is true also of an infectious disease. If there is an animal reservoir from which the infectious agent cannot be eliminated, it is sometimes possible, by eradicating the vector responsible for animal-to-man transmission, to prevent the introduction of the agent into human communities even though these communities are situated within or close to the endemic zoonotic zones. Some examples of highly successful eradication programs are described below:

The Texas fever tick, *Boophilus annulatus*, once caused yearly losses to cattlemen in the United States of $40 million (MacKellar 1942). An eradication program based on dipping of all cattle, horses, and mules in the infested areas was begun in 1906, and by 1943 all areas had been cleared of the tick except for a zone along the Rio Grande in Texas. From time to time, infestations of the tropical cattle tick, *B. microplus*, have been found in Florida. Cole and MacKellar (1956) state that "The total cost to Federal, State, and county governments of their cooperative conquest in eliminating this parasite has amounted to little more than the toll taken from the South in a single year by the fever tick before eradication work was started.

The screwworm fly, *Callitroga hominivorax*, is an obligate parasite in the larval stage, causing wound myiasis in man and animals. It was responsible for a $20-million loss annually to the livestock industry (Eddy and Bushland 1956). The eradication of this pest from Curaçao and the southeastern United States is a classic case of control by release of sterilized males. Setbacks in the program in the southwestern United States should not obscure the marked success in reducing losses of cattle and other livestock.

The first worldwide attempt at eradication of a disease was that of yellow fever, which was based on the belief that the disease was endemic only in tropical cities with a large *Aedes aegypti* population and supply of persons with viremia and of nonimmune infants or immigrants. By means of control of the urban vector, the disease was eliminated from the tropical cities in which it was endemic but was later found to persist in monkey reservoirs in the tropical rain forests of Africa and the Americas. Although the effort cost $14 million and caused 6 accidental deaths (Soper 1960), it was one of the greatest achievements in medicine.

Because yellow fever is transmitted in cities by *Aedes aegypti*, it was proposed in 1934 that the cooperative Yellow Fever Service should attempt to eradicate this species from Brazil. This appeared to have been accomplished by 1939. Because the problem of reinfestation of Brazil from neighboring areas was acute, the Pan American Health Organization in 1947 assumed the responsibility of coordinating the work in several countries. Eradication has progressed through much of tropical America, although reinfestations have occurred and the attempted eradication program in the United States has been abandoned.

The eradication of introduced *Anopheles gambiae* from Brazil in 1950 and from upper Egypt in 1945 were brilliant achievements. Soper and Wilson (1942) refer to a total budget of $1,225,000 for the anti-*gambiae* work in Brazil from October 1938 through 1940. A conservative estimate was that there were over 100,000 cases and 14,000 deaths during the malaria epidemic in Brazil caused by *A. gambiae* in 1938. Had *A. gambiae* been allowed to continue its depredations in South America, the costs in health and economic depression would have been incalculable.

The confident expectations expressed by the World Health Assembly in 1955 in its resolution to support malaria eradication programs throughout the world have not been achieved; however, the great success of these programs is apparent. The WHO Expert Committee on Malaria in its

Thirteenth Report states: "To have progressed this far and to have brought a sustained measure of well-being to a total of 953 million people, more than one quarter of the world's population, is an international achievement without parallel in the provision of public health service." This was possible because of the unique insecticidal properties of DDT.

For various reasons, there are problem areas where transmission cannot be completely prevented with intradomiciliary residual insecticides, and here it may be necessary to revert to control through species sanitation.

The choice between eradication and control must be dependent upon the evaluation of many factors, including the target species and accessibility of its breeding habitats; the extent of the area of infestation and the ease with which the area can be protected from reinfestation; the importance of the problem to the local inhabitants and governmental authorities; and whether full support will be given to the program over a reasonable length of time. The initial expenses are high, but if the campaign is successful, the problem has been eliminated and there should be no further costs, as in the case of *A. gambiae*. One has only to think of what the cost might have been in lives, morbidity, and attempted control if *A. gambiae* had been allowed to survive and to do in tropical America what it does in Africa, where intradomiciliary residual sprays have not brought about desired results. However, if eradication is achieved only in a limited geographic area, the necessity for continued surveillance becomes paramount.

The assumption that eradication is better than containment in the long run because a period of high costs is followed by little or no expenditures is challenged by Cohn (1973). In analyzing the costs of India's pre-eradication and eradication programs, Cohn concludes that at a discount rate of 12 to 15 percent there is no marked cost difference between control and eradication. What the eradication program did achieve was a reduction of India's malaria load from an estimated annual incidence of 75 million cases prior to 1953 to a reported 350,000 cases in 1969.* There probably will be no further economic or

*According to Scholtens et al. (1972), the numbers of positive cases in India, as determined by examination of blood films, increased to 694,647 and 1,323,118 in 1970 and 1971, respectively (see also Johnson 1973). The WHO *Weekly Epidemiological Record* No. 34 of August 24, 1973, reports a total of 1,362,806 positive blood smears for 1972.

sociological benefits by eliminating this small residue of malaria, so it now becomes necessary to apply the principle of discount cost analyses to determine whether containment or eradication would be the most advantageous. The problem is to maintain the gains already realized.

The organization of an eradication campaign is a gamble, and all concerned, especially those authorizing the expenditure of funds, should realize that progress may be slow, expensive, and often disappointing. Perhaps a serious error in the malaria eradication program was the promise an attack phase of about 5 years would be sufficient. This was based on the belief that almost all that was necessary was to make blanket applications of DDT to all houses. This was found not to be the case, and it has become obvious that in malaria as well as in other problems the strategy of eradication depends upon an intimate knowledge of the distribution, host relationships, and ecology of the target species. This is no less true if the objective is containment.

CONCLUSIONS

It is clear that a wide variety of tactics and strategies other than pesticides are available for the control of public health pests. The effectiveness and usefulness of some of these in control are well-established and scientifically validated. The procedures of source reduction through environmental sanitation, the utilization and management of larvivorous fish, and use of screening and repellents are such established alternatives to pesticides. Other alternatives are more limited in their current usefulness, and some are only in the earliest stages of evaluation. Some of these show great promise, and the potential and limitations of each have been discussed in the preceding sections.

Another most critical consideration will be how these several tactics will be used together in pest control or, better, in a pest management system. Identifying a new technology or tactic for pest control is not necessarily the same as identifying a functional alternative control. Such a technology must be made a part of a strategy of pest control and fitted into a management system. This adaptation of a technology into a system must include an economic analysis of that technology, including a benefit/cost study of the particular technology and of the modified system as an operational unit.

It is the view of the study team that no alternative to the use of pesticides is on the horizon that will radically change public health pest control strategy in the near future. Rather, the study team anticipates a gradual return to a more balanced, integrated approach utilizing a variety of pest control tactics appropriate to the ecology and behavior of the pest species, the epidemiology of the disease, and concern for environmental values. There will be an increase in reliance on source reduction, and the goal will be containment and reduction of disease rather than eradication.

REFERENCES

Anderson, C. H., and W. J. Schulte (1970) Conversion of thermal foggers for ULV application. Mosq. News 30: 209-212.

Aquatic Plant and Insect Control Committee, Pacific Northwest River Basins Commission, Water Resources Council (1968) Report on a Study of Mosquito Problems Associated with Development of the Crooked River Irrigation Project, Central Oregon, 1960-1966.

Bay, E. C. (1967) Mosquito control by fish: A present-day appraisal. WHO Chron. 21(10):415-423.

Bay, E. C. (1973) Exotic fish introductions for mosquito control: Possible and purported consequences. WHO mimeo VBC/WP/73.7.

Bay, E. C. (1974) Predator-prey relationships among aquatic insects. Annu. Rev. Entomol. 19:441-451.

Boike, A. H., and C. B. Rathburn (1973) An evaluation of several toxic rubber compounds as mosquito larvicides. Mosq. News 33:501-505.

Brown, A. W. A. (1972) Alternative methods of vector control, pp. 59-63. *In* Vector Control and the Recrudescence of Vector-Borne Diseases. Pan American Health Organization Sci. Publ. 238.

Brown, A. W. A., and A. G. Carmichael (1961) Lysine and alanine as mosquito attractants. J. Econ. Entomol. 54: 317-324.

Carlson, D. A., N. Smith, H. K. Gouck, and D. R. Godwin (1973) Yellow-fever mosquitoes: Compounds related to lactic acid that attract females. J. Econ. Entomol. 66:329-331.

Carmichael, G. T. (1972) Anopheline control through water management. Am. J. Trop. Med. Hyg. 21:782-786.

Chamberlain, N. P. (1929) Twenty-five years of medical activity on the Isthmus of Panama, 1904-1929. Mount Hope, Canal Zone: The Panama Canal Press.

Chapman, H. C. (1974) Biological control of mosquito larvae. Annu. Rev. Entomol. 19:33-59.

Chapman, H. C., T. B. Clark, and J. J. Petersen (1970) Protozoans, nematodes, and viruses of anophelines. Misc. Publ. Entomol. Soc. Am. 7:134-139.

Clark, T. B., H. C. Chapman, and T. Fukuda (1969) Nuclear polyhedrosis and cytoplasmic polyhedrosis virus infections in Louisiana mosquitoes. J. Invert. Pathol. 14: 284-286.

Cline, R. E., and M. H. Hall (1973) Larvicidal activities of N,N'-dialkylalkylenediamines and other fatty diamines in oil films and water dispersions. J. Econ. Entomol. 66:697-702.

Cline, R. E., D. P. Wilton, and R. W. Fay (1969) Aliphatic amines ovicidal to the yellow-fever mosquito. J. Econ. Entomol. 62:981-986.

Cohn, E. J. (1973) Assessing the costs and benefits of anti-malaria programs: The Indian experience. Am. J. Public Health 63:1086-1096.

Cole, T. W., and W. M. MacKellar (1956) Cattle tick fever, pp. 310-313. *In* Animal Diseases, Yearbook of Agriculture, 1956. Washington, D.C.: U.S. Government Printing Office.

Corbet, P. S., and A. Griffiths (1963) Observations on the aquatic states of two species of *Toxorhynchites* in Uganda. Proc. Roy. Entomol. Soc. London (A) 38:125-135.

Couch, J. N. (1972) *Coelomomyces*: Mass Production of a fungus that kills mosquitoes. Proc. Nat. Acad. Sci. U.S. 69:2043-2047.

Couch, J. N., and C. J. Umphlett (1963) Coelomomyces infections, pp. 149-188. *In* Vol. 2, E. A. Steinhause (ed.). Insect Pathology. New York: Academic Press.

Covell, C. V., Jr., and V. H. Resh (1971) Relative abundance of *Culex pipiens* and *Culex restuans* in catch basins in Jefferson County, Kentucky. Mosq. News 31: 73-76.

Critcher, T. S. (1970) North Carolina Wildlife Resources Commission, pp. 101-108. *In* Proc. Workshop on Mosquito Control in North Carolina. N.C. State Univ.

Davidson, G., J. A. Odetoyinbo, B. Colussa, and J. Coz (1970) A field attempt to assess the mating competitiveness of sterile males produced by crossing two member species of the *Anopheles gambiae* complex. Bull. WHO 42:55-67.

De Bach, P. (ed.) (1964) Biological Control of Insect Pests and Weeds. London: Chapman & Hall.

De Bach, P. (1966) The competitive displacement and coexistence principles. Annu. Rev. Entomol. 11:183-212.

Dunn, R. L., and F. E. Strong (1973) Control of catch-basin mosquitoes using Zoecon ZR 515 formulated in the slow release polymers. Mosq. News 33:110-111.

Eddy, G. W., and R. C. Bushland (1956) Screwworms that attack livestock, pp. 172-175. *In* Yearbook of Agriculture, 1956. Washington, D.C.: U.S. Government Printing Office.

FAO/WHO (1973) The Use of Viruses for the Control of Insect Pests and Disease Vectors. Report of a joint FAO/WHO Meeting on Insect Virus. Geneva, 22-27 November 1972. FAO Agr. Studies No. 91.

Gartrell, R. E., W. W. Barnes, and G. S. Christophers (1972) Environmental impact and mosquito control water resource management projects. Mosq. News 32:337-343.

Gerberich, J. B., and M. Laird (1968) Bibliography of papers relating to the control of mosquitoes by the use of fish (an annotated bibliography for the years 1901-1966). FAO Fish. Tech. Paper 75:1-70.

Gilbert, I. H., H. K. Gouck, and N. C. Smith (1957) New insect repellent. Soap Cosmet. Chem. Spec. 33:115-117, 129-133, 95-99, 109.

Gouck, H. K., D. R. Goodwin, K. Posey, C. E. Schreck, and D. E. Weidhaas (1971) Resistance to aging and rain of repellent-treated netting used against salt-marsh mosquitoes in the field. Mosq. News 31:95-99.

Greenfield, H. R. (1971) Acceptance of responsibility for area-wide drainage in an agricultural area, pp. 93-94. *In* Proc. Pap. 39th Annu. Conf. Calif. Mosq. Cont. Assoc. Turlock, California: California Mosquito Control Association.

Gubler, D. J. (1970) Competitive displacement of *Aedes (Stegomyia) polynesiensis* Marks by *Aedes (Stegomyia) albopictus* Skuse in laboratory populations. J. Med. Entomol. 7:229-235.

Hackett, L. W. (1937) Malaria in Europe. London: Oxford University Press.

Harrington, R. W., and E. S. Harrington (1961) Food selection among fishes invading a high subtropical salt marsh: from onset of flooding through the progress of a mosquito brood. Ecology 42(4):646-666.

Hayden, D. L., J. A. Mulrennan, W. V. Weeks, and F. M. Ulmer (1973) Ultra low volume dispersal equipment for military helicopters. Mosq. News 33:387-391.

Heimpel, A. M. (1967) A critical review of *Bacillus thuringiensis* var. *thuringiensis* Berliner and other crystalliferous bacteria. Annu. Rev. Entomol. 12:287-322.

Hildemann, W. H., and R. L. Walford (1963) Annual fishes: Promising species as biological control agents. J. Trop. Med. Hyg. 66:163-166.

Hinman, E. H. (1941) The management of water for malaria control, pp. 324-332. *In* Symposium on Human Malaria. AAAS Publ. No. 15.

Hiscox, K. J., E. J. Kingford, and W. E. Hazeltine (1974) Annual fish of the genus *Cynolebias* for mosquito control. Proc. Calif. Mosq. Control Assoc.

Hoy, J. B., and J. J. Petersen (1973) Fish and nematodes--current status of mosquito control techniques. Proc. Calif. Mosq. Control Assoc. 41:49-50.

Ikeshoji, T., and M. S. Mulla (1970a) Overcrowding factors of mosquito larvae. J. Econ. Entomol. 63:90-96.

Ikeshoji, T., and M. S. Mulla (1970b) Growth retarding and bacteriostatic effects of the overcrowding factors of mosquito larvae. J. Econ. Entomol. 63:1737-1743.

Innes, W. T. (1917) Goldfish Varieties and Tropical Aquarium Fishes. Philadelphia: Innes & Sons.

Jakob, W. L. (1973) Developmental inhibition of mosquitoes and the house fly by urea analogues. J. Med. Entomol. 10:452-455.

Jakob, W. L., R. W. Fay, and D. P. Wilton (1970) Field trials of an amine ovicide against *Aedes aegypti* (L.). Mosq. News 30:191-194.

Jenkins, D. W. (1964) Pathogens, Parasites, and Predators of Medically Important Arthropods: Annotated List and Bibliography. Supplement to Bull. WHO Vol. 30.

Johnson, D. R. (1973) Recent developments in mosquito-borne diseases: Malaria. Mosq. News 33:341-347.

Junkert, B., and A. P. Quintana (1972) Economic considerations in pasture rehabilitation, pp. 35-37. *In* Proc. Pap. 40th Annu. Conf. Calif. Mosq. Cont. Assoc. Turlock, California: California Mosquito Control Association.

Kilpatrick, J. W., D. A. Eliason, and M. F. Babbitt (1970) Studies of the potential effectiveness of ultra low volume aerial applications of insecticides against *Aedes aegypti* larvae. Mosq. News 30:250-258.

Kistner, T. P., and F. A. Hayes (1970) White-tailed deer as hosts of cattle fever ticks. J. Wildl. Dis. 6:437-440.

Kliewer, J. W., T. Miura, R. C. Husbands, and C. H. Hurst (1966) Sex pheromones and mating behavior of *Culiseta inornata*. Ann. Entomol. Soc. Am. 59:530-533.

Kumm, H. W. (1941) The adaptability of control measures to malaria vectors of the Caribbean Region, pp. 359-364. *In* Symposium on Human Malaria. AAAS Publ. No. 15.

Washington, D.C.: American Association for the Advancement of Science.

Laird, M. (1967) A coral island experiment: a new approach to mosquito control. WHO Chron. 21:18-26.

Laird, M. (1970) Microbial control of arthropods of medical importance, pp. 387-406. *In* Microbial Control of Insects and Mites, H. D. Burges (ed.). New York: Academic Press.

Laird, M. (1971) The biological control of vectors. Science 11:590-592.

Laven, H. (1967) Eradication of *Culex pipiens fatigans* through cytoplasmic incompatibility. Nature 216:383-384.

Laven, H., J. Cousserans, and G. Guille (1971) Inherited semisterility for control of harmful insects. III. A first field experiment. Experientia 29:1355-1357.

Lewis, L. F., and D. M. Christenson (1971) Solution to a drainage problem in a simulated pasture with implications for its application in actual pastures, pp. 72-73. *In* Proc. Pap. 39th Annu. Conf. Calif. Mosq. Cont. Assoc. Turlock, California: California Mosquito Control Association.

Lewis, L. F., D. M. Christenson, and W. M. Rogoff (1972) Vertical drainage for mosquito abatement in small problem areas associated with irrigated fields: A progress report, pp. 61-63. *In* Proc. Pap. 40th Annu. Conf. Calif. Mosq. Cont. Assoc. Turlock, California: California Mosquito Control Association.

Lowrie, R. C., Jr. (1973) Displacement of *Aedes* (*S.*) *polynesiensis* Marks by *A.* (*S.*) *albopictus* Skuse through competition in the larval stage under laboratory conditions. J. Med. Entomol. 10:131-136.

MacKellar, W. M. (1942) Cattle tick fever, pp. 572-578. *In* Yearbook of Agriculture, 1942. Washington, D.C.: U.S. Government Printing Office.

MacLaren, J. P. (1972) A brief history of sanitation in the Canal Zone, 1913-1972. Div. Sanit., Health Bureau. Canal Zone Government.

Madelin, M. F. (1966) Fungal parasites of insects. Annu. Rev. Entomol. 11:423-448.

Mayor's Advisory Committee on Mosquito Control, City of New Orleans (1970) Annual Report.

McCabe, E. T., W. F. Barthel, S. I. Gertler, and S. A. Hall (1954) Insect repellents. III. N,N-Diethylamides. J. Org. Chem. 19:493-498.

Miller, T. A., M. A. Lawson, L. L. Nelson, and W. W. Young (1973) Laboratory and field evaluation of encapsulated formulations of malathion and Abate. Mosq. News 33:413-417.

Mount, G. A., M. V. Meisch, J. T. Lee, N. W. Peirce, and K. F. Baldwin (1972) Ultra low volume ground aerosols of insecticides for control of rice field mosquitoes in Arkansas. Mosq. News 32:444-446.

Mulhern, T. D. (1973a) The third era of mosquito control in California has just begun. Paper presented at the 1973 Annual Meeting, California Mosquito Control Association.

Mulhern, T. D. (1973b) Comparisons of structure and approach to control in U.S. Mosquito Abatement programs. Future mosquito abatement in the United States. Unpublished paper presented at meeting of Public Health Study Team, Berkeley, California, September 26-27.

Mulla, M. S. (1967) Biocidal and biostatic activity of aliphatic amines against southern house mosquito larvae and pupae. J. Econ. Entomol. 60:515-522.

Mulla, M. S., H. A. Darwazch, and P. A. Gillies (1970) Evaluation of aliphatic amines against larvae and pupae of mosquitoes. J. Econ. Entomol. 63:1472-1475.

Myers, G. S. (1965) *Gambusia*, the fish destroyer. Trop. Fish Hobby., January 1965, pp. 31-32, 53-54.

Myers, G. S. (1973) WHO conference on the safety of biological agents for arthropod control. WHO mimeo WHO/VBC/73.445 p. 65.

Nakagawa, P. Y., and J. Ikeda (1969) Biological control of mosquitoes with larvivorous fish in Hawaii. Unpublished document. WHO/VBC/69.173.

National Academy of Sciences (1973) Mosquito control: Some perspectives for developing countries. Washington, D.C.

Nave Rebollo, O., E. Parada S., and A. Guerra (1973) El paludismo en El Salvador. Análisis de las campañas de control y erradicación. Rev. Inst. Invest. Med. 2:3-30.

Nelson, L. L., T. A. Miller, J. T. Whitlaw, W. W. Young, and M. A. Lawson (1973) Laboratory and field evaluation of selected polyethylene formulations of chlorpyrifos. Mosq. News 33:401-412.

Nijhout, H. F., and G. B. Craig (1971) Reproductive isolation in *Stegomyia* mosquitoes. III. Evidence for a sexual pheromone. Entomol. Exp. Appl. 14:399-412.

Oates, N. K. (1970) Present approaches to mosquito control in North Carolina. Proc. Workshop in Mosquito Control in North Carolina. N.C. State University: 15-24.

Pal, R., and L. E. LaChance (1974) The operational feasibility of genetic methods for the control of insects of medical and veterinary importance. Annu. Rev. Entomol. 19:269-291.

Patterson, R. S., D. E. Weidhaas, H. R. Ford, and C. S. Lofgren (1970) Suppression and elimination of an island population of *Culex pipiens quinquefasciatus* with sterile males. Science 168:1368-1370.

Peters, R. H. (1971) Reclamation of overflow lands to agriculture by a mosquito abatement district as a means of permanently eliminating an intolerable mosquito pest. Proc. Pap. 39th Ann. Conf. Calif. Mosq. Cont. Assoc.: 97-98.

Rai, K. S., K. K. Grover, and S. G. Suguna (1973) Genetic manipulation of *Aedes aegypti*: incorporation and maintenance of a genetic marker and a chromosomal translocation in natural populations. Bull. WHO 48:49-56.

Rajagopalan, P. K., M. Yasuno, and G. C. LaBrecque (1973) Dispersal and survival in the field of chemosterilized, irradiated and cytoplasmically incompatible male *Culex pipiens fatigans*. Bull. WHO 48:631-635.

Reeves, E. L., and C. Garcia (1971) Pathogenicity of bicrystalliferous *Bacillus* isolate for *Aedes aegypti* and other Aedine mosquito larvae. Proc. IV Int. Colloq. Insect. Pathol., College Park, Maryland, August 1970. pp. 219-228.

Reynolds, D. G. (1972) Experimental introduction of a microsporidian into a wild population of *Culex pipiens fatigans* Wied. Bull. WHO 46:807-812.

Robinson, E. C. (1971) A long-term source reduction program, now in the maintenance phase. Proc. Pap. 39th Annu. Conf. Calif. Mosq. Cont. Assoc.: 92-93.

Robinson, T. (1907) Panama 1861-1907. New York: Star and Herald Company.

Rogers, A. J., and C. B. Rathburn (1960) Improved methods of formulating Paris green larvicide. Mosq. News 20: 11-14.

Sacher, R. M. (1971) A mosquito larvicide with favorable environmental properties. Mosq. News 31:513-516.

Schaefer, C. H., and W. H. Wilder (1972) Insect development inhibitors: A practical evaluation of mosquito control agents. J. Econ. Entomol. 65:1066-1071.

Schaefer, C. H., and W. H. Wilder (1973) Insect development inhibitors. 2. Effects on target mosquito species. J. Econ. Entomol. 66:913-916.

Scholtens, R. G., R. L. Kaiser, and A. D. Langmiur (1972) An epidemiologic examination of the strategy of malaria eradication. Int. J. Epidemiol. 1:15-24.

Sharma, V. P., R. S. Patterson, K. K. Grover, and G. C. LaBrecque (1973) Chemosterilization of the tropical house mosquito *Culex pipiens fatigans*: Laboratory and field cage studies. Bull. WHO 48:45-48.

Sholdt, L. L., D. A. Ehrhardt, and A. G. Michael (1972) A guide to the use of the mosquito fish, *Gambusia affinis*, for mosquito control. U.S. Navy Environ. Prev. Med. Unit No. 2. EPMU2 PUB6250/3 (2-72).

Simmons, J. S. (1939) Malaria in Panama. Am. J. Hyg. Monogr. Ser. No. 13.

Smith, C. N. (1972) Genetic control of mosquitoes: program of the WHO/ICMR collaborative research unit. Indian Counc. Med. Res. Tech. Rep. Ser. 20:129-134.

Smith, C. N., N. Smith, H. K. Gouck, D. E. Weidhaas, I. H. Gilbert, M. S. Mayer, J. B. Smittle, and A. Hofbauer (1970) L-Lactic acid as a factor in the attraction of *Aedes aegypti* to human hosts. Annu. Entomol. Soc. Am. 63:760-770.

Smith, R. F. (1973) Considerations on the safety of certain biological agents for arthropod control. Bull. WHO 48:685-698.

Soper, F. (1960) Eradication versus control in communicable disease prevention. J. Am. Vet. Med. Assoc. 137: 234-238. See also Soper, F. L., "Building the Health Bridge." Bloomington: Indiana University Press, 1970. 330-336.

Soper, F. L., and D. B. Wilson (1942) Species eradication: A practical goal of species reduction in the control of mosquito-borne disease. J. Nat. Malar. Soc. 1:5-24.

Spielman, A. (1972) Synthetic juvenile hormones as larvicides for mosquitoes. Ind. Trop. Health 7:67-69.

Steffan, W. A. (1970) Hawaiian toxorhynchites (Diptera: Culicidae). Proc. Hawaiian Entomol. Soc. 20(1):141-155.

Tennessee Valley Authority (1973) Environmental Statement. Vector Control Program. Chattanooga, Tennessee.

Trpis, M. (1973) Predator-prey oscillations in density of *Toxorhynchites brevipalpus* and *Aedes aegypti* in a suburban habitat in East Africa. Bull. WHO. (In press.)

Valdivieso, J. A. (1950) The struggle against malaria. Phase I up to 1946. Read at 2nd Natl. Health Congr. of El Salvador.

Vannote, R. L. (1971) Current mosquito control problems in New Jersey, pp. 41-42. *In* Proc. Pap. 39th Annu. Congr. Calif. Mosq. Cont. Assoc. Turlock, California: California Mosquito Control Association.

Weidhaas, D. E., M. C. Bowman, G. A. Mount, C. S. Lofgren, and H. R. Ford (1970) Relationship of minimum lethal dose to the optimum size of droplets of insecticides for mosquito control. Mosq. News 30:195-200.

Welch, H. E. (1962) Nematodes as agents for insect control. Proc. Entomol. Soc. Ontario 92:11-19.

Whitten, G. R. (1969) Report on Rockefeller Drainage Project.

Whitten, M. J., and R. Pal (1974) The Use of Genetics in Insect Control. New York: Elsevier.

WHO (1973) Conference on the Safety of Biological Agents for Arthropod Control. WHO mimeo WHO/VBC/73.445.

Wright, J. W., R. F. Fritz, and J. Haworth (1972) Changing concepts of vector control in malaria eradication. Annu. Rev. Entomol. 17:75-102.

5

DIFFUSION OF MORE INTENSIVE AGRICULTURAL TECHNOLOGIES AND THEIR CONSEQUENCES

With the greatly expanded introduction of technology from the developed northern temperate countries into tropical environments after World War II, increasing concern has been expressed over the undesirable features of this transfer of technology (Farvar and Milton 1973). With this in mind, the study team has considered the impact of the introduction of new agricultural technologies on the problems of public health. As this is such a broad area of interaction, the study team concentrated on the impact of the changing technology of rice production in Asia (particularly Southeast Asia) on public health problems (see Appendix A). We also consider the consequences of pesticide use in their direct (intoxication and mortality from pesticides) and indirect (induction of pesticide resistance in disease vectors) impact on public health. In this case, we use the situation in Central America as our primary example.

There are over a hundred developing countries, and they contain 70 percent of the world's population and occupy two-thirds of the earth's land surface. In attempts to transfer the technology and other advantages of the developed countries to these developing countries, a number of ecological, sociological, political, and economic mistakes have been made. Nevertheless, on balance the gains have been significant in the developing countries. "During the past decade their economies grew at an average rate of 5.6 percent per year. In the last 10 years they have increased manufacturing output by 92 percent. Food production has increased by 36 percent. Average life expectancy reached 57 years in 1971. In the last decade, the developing countries have progressed faster than the developed countries did at comparable time periods in their history" (Hannah 1973). Nevertheless, because of population growth, per capita food production is barely higher than it was 10 years ago, and mass poverty prevails.

The world has seen in recent years an amazing change in the race between food production and human population increase. Views differ on where we stand today in this race. Nevertheless, tremendous gains in food production have occurred in many parts of the world, and this trend is expected to continue. This widely publicized phenomenon--often termed the Green Revolution, though in fact it is more evolutionary than revolutionary--has resulted from a combination of many factors, chief among which are (a) the introduction of new high-yielding crop varieties; (b) the availability of purchased production inputs, e.g., fertilizers, pesticides, tractors; (c) new crop management technology (including double and multiple cropping); (d) improved irrigation capability; and (e) the long-term cumulative effect of development efforts by national governments and international agencies. It should, of course, be recognized that part of the gains in food production in some years has also been the result of favorable weather.

Pressured by a multitude of pests over many centuries, man's crop plants have become adapted through natural selection to the pressures of agroecosystems. This state is stabilized by an array of genetic factors for high yield combined with tolerance to low fertility, pest attack, and other environmental stresses.

In contrast to traditional agriculture, modern agriculture is a more intensified system that integrates capital inputs with management technology to maximize production per unit of area at minimum cost per unit of production, ideally on a continuing basis. Many of the practices developed to achieve this goal contribute significantly to increased plant pest problems and thus may actually prevent achievement of the goal. For example, plant introduction and exchange has resulted in varieties with higher yields, resistance to pests, and other desirable qualities; but this plant movement may carry with it new pests and disease pathogens, and the introduced plant types may be susceptible to indigenous pests and diseases.

Many cultural practices of modern agriculture may enhance susceptibility to disease or attack by insects (Apple 1972). These include (a) fertilization, which produces larger and more succulent plants that are sometimes more susceptible to disease or insect damage than plants grown at lower nutritional levels; (b) irrigation, which, unlike the fluctuating soil moisture levels under natural rainfall conditions, favors many disease and insect pests; (c) tillage and other soil manipulations, which may increase the incidence of disease over that experienced in no-tillage or limited-tillage cultures; (d) double and triple cropping, which promotes rapid increase of pest populations; and (e) denser plant populations, with resulting microenvironmental changes that favor the development of some pests. These same cultural practices may at times inhibit some pests, but, in general, the balance is one favoring increased pest and disease incidence.

The Green Revolution has intensified the introduction of new practices into a sizable number of developing nations. Although "revolution" may overstate the rapidity with which these practices have been adopted, the increased production that resulted has nevertheless been surprising and gratifying (Castillo 1970). Motivated by the increased production with the new practices, many developing countries and international organizations have placed increased emphasis on the development and introduction of new agricultural technology.

These modernizing practices, which also enhance the potential for destructive pest attacks, are being introduced with only limited attention to crop protection as a component of agricultural development programs. We do not question the validity of these developments--there is now no other alternative. The fact remains that the changed agroecosystems resulting from the introduction of new methodologies produce shifts in, and very often an intensification of, pest and disease problems. This proven hazard is not today properly reflected in most of the development programs around the world. There is mounting evidence indicating that pest and disease problems in the developing countries are becoming more severe, indeed in some cases devastating, as modern practices are introduced.

Unless bold measures are taken to protect the food crops of developing nations against the ravages of pests and diseases, the production gains recently realized could vanish, and hope for the future could be lost. Along with the introduction of new production technology, the introducers and the recipient developing nations must assure the development of an adequate crop protection response capability. This must involve significant effort in the training and retraining of crop protection and pest management specialists, the organization of new types of programs for research at the adaptive and implementation levels, and the education of the general public and farmers as to the significance of crop protection to their welfare.

Minimizing the disruptive effects of pest control activities on environmental quality can contribute to an enhanced environment and at the same time to the improved nutrition and health of man in all parts of the world.

The insertion of a chemical pesticide into an agroecosystem has as its objective a change in the living conditions of at least two components of the system. Chain reactions of enormous complexity may be set in motion by the application of biologically active pesticides, such as the organochlorine, organophosphorus, and carbamate compounds used during the last quarter-century. The environmental disruptions resulting from use of the synthetic organic pesticides have resulted in substantial alteration of the faunal composition of our agroecosystems, especially those that have received intensive treatment--deciduous fruits and cotton, for example. Some of the changes that have been observed most frequently following applications of pesticides have been severe outbreaks of secondary pests and of normally minor species and the rapid resurgence of treated populations.

Considerable field experimentation in recent years has produced strong evidence to support the proposition that natural enemy suppression by pesticides is a major cause of change in pest status and resurgence of treated populations. Nevertheless, some evidence is available and shows that factors other than the destruction of natural enemies are also involved in these population phenomena. A better understanding than now exists of the complex interactions taking place in agricultural ecosystems is needed. The interactions between pests and their hosts, natural enemies, and competitors and between the natural enemies and their natural enemies and competitors must be better understood in order to develop effective pest management systems.

Unless a broad ecological approach such as that emphasized by integrated-control-type programs is initiated, additional pesticide abuse situations will arise. Complete dependence on hazardous, broad-spectrum pesticides over a period of time not only fails to control the pests in question but actually aggravates pest problems, endangers human health, and threatens environmental quality. Furthermore, pesticide misuse imposes an additional real cost on food production.

The integrated control strategy employs the idea of maximizing natural control forces and utilizes other pest management tactics with a minimum of environmental disturbance and only when threatened crop losses justify action. Use of natural enemies and plant resistance are basically compatible and supportive in the integrated control strategy. Cultural control, a third basically compatible tactic, is commonly used in ways to expose the pests to adverse weather, to disrupt their natural development, to increase the action of natural enemies, or to increase the crop's resistance. Chemicals, although not always compatible with the use of natural enemies, can often furnish a reliable immediate solution to a problem. Thus, pesticides are an important and necessary element in integrated control programs. Finally, a basic fund of ecological and biological knowledge is needed to guide decision making in the integrated control strategy.

In the utilization of natural enemies, consideration must go far beyond the techniques of classical biological control, i.e., introduction of parasites and predators into new areas. Attention should also be given to assessing and understanding the role of natural biological controls, manipulation of the environment to increase efficiency of existing natural enemies, periodic colonization of natural enemies, supplemental feeding of natural enemies, and utilization of the invertebrate pathogens such as viruses.

The possibilities suggested by introduction of new invertebrate pathogens and utilization of indigenous ones, such as insect viruses, have hardly been approached. It is probable that pathogens selective for certain species or groups of insects and innocuous to vertebrates abound in nature, but too little effort has been made to find them, characterize them, and develop them for practical use. Pathogens have many of the advantages of chemical insecticides and they lack many of their disadvantages; furthermore, the available pesticide application technology is adaptable to them. Some of the known pathogens are quickly and highly effective, specific in activity, safe and biodegradable. In some cases, they can be readily stored. Their cost, lack of proved reliability, questions of patentability, and problems of registration clearance are some of their disadvantages and barriers to their development.

Chemical pesticides remain in many situations a most powerful and dependable tool for the management of pest populations. They can be more effective, dependable, economical, and adaptable for use in a wide variety of situations than many other proven tools for maintaining pest populations at subeconomic levels. Indeed, use of chemical pesticides is the only known method for control of many of the world's most important pests of agriculture and public health importance. No other tool is so easy to manipulate, and none can be brought to bear so quickly on outbreak populations.

Narrowly selective chemicals appear to offer an almost ideal means of pest control. However, only a very few such chemicals have been discovered and developed for commercial use. Future prospects for additional developed chemicals have become very dim. Historically, the chemical industry, for the most part, has had little interest in finding and developing this type of compound. The financial return on investments in research and development of truly physiologically selective insecticides is small when compared with that for the broad-spectrum compounds now so widely used for insect control. Except for a relatively few key pests of major crops--e.g., boll weevil, rice stem borer, and codling moth--the chemical industry would be hard pressed to recover research and development costs of monotoxic compounds for pest control.

Fortunately, it is not always necessary to rely upon the physiological selectivity of chemicals to obtain some of the specific effects required in integrated control and other pest management systems. Ecological selectivity obtained by the discriminating use of even the broadest-spectrum insecticides can be employed in many cases for the development of effective, economical, and ecologically sound pest control programs. Development of such programs is presently limited to some extent by a lack of knowledge of the ecology, biology, and behavior of pest/natural-enemy/crop complexes. A more serious limiting factor is a shortage of properly trained, imaginative, and capable applied crop protection specialists dedicated to the development of pest management systems based on the principles of integrated control.

While it is recognized that pesticide chemicals have been and will continue to be an essential part of crop protection, current practices in pesticide use have not always been sound, not only in terms of food production, but also from the standpoint of human health and environmental quality. There is substantial need for trained personnel to assist and guide the pesticide management process--i.e., the proper selection, procurement, formulation, packaging, shipment, storage, marketing, application, and disposal of pesticides.

PATTERNS OF PESTICIDE USE

World Pesticide Production and Use

There are no official and complete statistics on worldwide pesticide production and use. The Food and Agricultural Organization (FAO) of the United Nations publishes in its annual *Production Yearbook* data on pesticides "used in, or sold to agriculture" in a number of countries, by quantities of active ingredients. Unfortunately, these data are very incomplete, since FAO relies on voluntary reports by its member nations. Up to the present, cooperation in this reporting has been much less than 50 percent. Thus, the FAO data are not sufficient for an estimate of the total world trade in pesticides (Table 5-1).

The agricultural chemical trade magazine *Farm Chemicals* has published a world pesticide market survey conducted by its staff. This survey, supplemented by estimates from industry sources and RvR Consultants, estimate the world market for agricultural pesticides in 1971 at $2.5 billion at the consumer price level (Table 5-2). Herbicide use on small grains in Europe and use of insecticides and fungicides on rice (especially in Japan) account for the majority of non-U.S. pesticide uses on grain crops. Food and feed grains in developing countries receive only rather small quantities of pesticides.

Fruit, vegetable, and horticultural crops account for $570 million worth of pesticides worldwide. The U.S. consumption for these crops totals $160 million worth of insecticides, fungicides, and herbicides. Other countries spent $210 million on fungicides, $150 million on insecticides, and $40 million on herbicides for use on these crops.

Pesticides used on all other crops (including soybeans, sugar beets, alfalfa, other field crops, and tobacco) combined and pesticides used on livestock account for the remaining expenditure on pesticides. U.S. domestic use of herbicide on soybeans represents a sizable share of this total. Again, the lion's share of the balance is used in other developed countries.

Pesticide Use in Developing Countries

There are no official U.S. or international statistics, or other reliable published reports, on the amounts of pesticides imported into, produced in, or used in developing countries. The data in Table 5-3 indicate that pesticide consumption in all developing countries combined accounts for a relatively small share of total world pesticide consumption, probably 15-20 percent. Unfortunately, it is not possible to account with greater accuracy for pesticide inputs into developing countries at present.

However, complete information is available on U.S. pesticide exports by quantities, by value, by destination and by individual pesticide chemicals or groups of chemicals. The value of U.S. pesticide exports has increased from about $100 million in 1961 to slightly over $250 million in 1971. Table 5-3 presents a detailed analysis of the value of U.S. pesticide exports by destination for

TABLE 5-1 World Markets for Agricultural Pesticides[a] and Cropland Areas, 1971[b]

Area	Pesticide Use[c] (thousands of dollars)	Pesticide Use[c] (%)	Cropland[d] (thousands of hectares)	Cropland[d] (%)	$ Pesticides per Hectare Cropland
United States	1,030,000	41	176,500	16	5.84
Canada	110,000	5	43,500	4	2.53
Europe	510,000	20	147,000[e]	14	3.47
Asia	450,000	18	351,000[f]	32	1.28
Latin America	205,000	10	120,000	11	2.08
All other regions[g]	150,000	6	252,000	23	0.60
TOTAL	2,500,000	100	1,090,000	100	2.29

[a] Excluding industrial, commercial, home, garden, lawn, and turf pesticides.
[b] Sources: Farm Chemicals, Jan. 1973, 27-33; Industry and RvR Consultants' Estimates; FAO Production Yearbook, Vol. 25, 1974.
[c] Basis: Consumer prices.
[d] Cropland defined as arable land and land under permanent crops, or total agricultural area exclusive of permanent meadows and pastures.
[e] Exclusive of Soviet Union.
[f] Exclusive of People's Republic of China.
[g] Africa, Australia, and Oceania.

TABLE 5-2 World and U.S. Markets for Agricultural Pesticides[a] by End Use, 1971[b,c]

Type of Use	World Market (thousands of dollars)	World Market (%)	U.S. Market (thousands of dollars)	U.S. Market (%)	U.S./World (%)
Herbicides	1,150,000	46	640,000	62	56
Insecticides	850,000	34	225,000	22	26
Fungicides	340,000	14	65,000	6	19
Other	160,000	6	100,000	10	63
TOTAL	2,500,000	100	1,030,000	100	41

[a]Excluding industrial, commercial, home, garden, lawn, and turf pesticide uses.
[b]Basis: Consumer prices.
[c]SOURCE: Farm Chemicals, Jan. 1973, 27-33; RvR Consultants' Estimates.

TABLE 5-3 U.S. Pesticide Exports by Destination: Total Exports by Value (in thousands of dollars)

Country	1966	1967	1968	1969	1970	1971
Africa						
UAR	11,794	10,968	9,661	3,261	2,248	2,870
Kenya	677	1,103	1,601	1,074	2,364	n.a.
Tunisia	n.a.	n.a.	n.a.	n.a.	25	105
Dahomey	n.a.	n.a.	n.a.	n.a.	--	--
Liberia	n.a.	n.a.	n.a.	n.a.	129	139
Ghana	n.a.	n.a.	n.a.	n.a.	55	29
Ethiopia	n.a.	n.a.	n.a.	n.a.	56	673
South Africa	4,423	5,211	5,985	5,986	7,113	8,751
All others	balance	balance	balance	balance	5,583	balance
Total Africa	20,737	21,899	22,038	14,237	17,573	19,952
Near and Central Asia						
Pakistan	5,960	7,343	11,850	7,773	13,080	2,033
India	5,852	5,612	11,128	2,527	2,974	7,557
Turkey	1,147	680	1,497	727	2,261	3,152
Sri Lanka	n.a.	n.a.	n.a.	n.a.	102	265
Far East Asia						
Philippines	n.a.	1,309	2,114	1,251	1,792	2,228
Thailand	1,582	3,471	2,422	4,403	2,525	2,444
S. Vietnam	n.a.	852	1,491	714	1,318	2,344
Indonesia	n.a.	n.a.	n.a.	n.a.	878	542

TABLE 5-3 (continued)

Country	1966	1967	1968	1969	1970	1971
S. Korea	n.a.	n.a.	n.a.	n.a.	1,068	1,532
All others	balance	balance	balance	balance	16,462	24,186
Total Asia	31,461	40,400	53,153	38,578	42,460	46,283
South America						
Brazil	8,969	6,046	9,393	8,661	10,918	10,487
Colombia	9,814	3,851	8,107	8,486	8,538	9,698
Venezuela	3,698	4,106	4,112	3,162	3,916	3,478
Argentina	1,579	2,302	2,511	2,523	3,376	3,122
Paraguay	n.a.	n.a.	n.a.	n.a.	210	374
Chile	n.a.	n.a.	n.a.	n.a.	833	331
All others	balance	balance	balance	balance	2,761	3,138
Total South America	28,543	20,497	28,383	27,020	30,552	30,628
Middle America						
Guatemala	n.a.	3,286	2,872	2,908	1,600	2,500
Nicaragua	5,420	8,952	6,234	3,478	3,315	3,792
Dom. Republic	n.a.	n.a.	n.a.	n.a.	1,605	1,464
Costa Rica	1,323	1,563	2,533	2,395	2,405	2,812
Honduras	1,348	1,918	2,307	2,352	2,080	2,412
Mexico	11,415	8,543	12,315	14,325	5,018	10,554
El Salvador	n.a.	3,560	3,519	3,689	1,511	3,202
All others	balance	5,387	6,791	5,238	4,474	3,958

Total Middle America	30,744	33,209	36,571	34,385	22,008	30,694
Canada	15,833	20,401	24,704	23,371	17,318	18,960
Total Americas	46,577	53,610	61,275	57,756	39,326	49,654
Europe	39,003	51,054	66,782	55,981	83,983	97,457
Australia and Oceania	7,812	8,273	9,848	9,249	6,419	7,628
Grand Total	173,633	195,723	241,479	202,821	220,313	251,602

SOURCE: USDA, Agr. Stabilization and Conservation Service.

the period 1966-1971. Selected developing countries that imported pesticides in 1970 or 1971 are shown individually in this table, allowing a country-by-country comparison. For a number of countries, for instance, the United Arab Republic, Pakistan, India, Thailand, Colombia, and Nicaragua, the table shows considerable variations from year to year in the total value of pesticides imported from the United States. This does not indicate corresponding ups and downs in total pesticide use in the countries concerned, but, rather, reflects changes in the sources of supply resulting from world competition in the pesticide market. For instance, the United States supplied major quantities of insecticides for use on cotton in Egypt in the mid-1960s, but was later replaced by other suppliers.

Table 5-4 summarizes total U.S. pesticide exports by products or groups of products, and by technical and formulated pesticides, for 1970 and 1971. The total quantity of insecticides exported was almost the same in 1971 as in 1970. The total value of insecticide exports increased from $119 million in 1970 to $137 million in 1971, indicating price increases and/or shifts to more expensive products. The volume and value of insecticides exported as technical active ingredient increased, while the volume and value of formulated insecticides declined. This is indicative of the continuing expansion of formulating capabilities in other countries, especially developing countries.

Exports of herbicides are on the increase, in line with the rapidly expanding worldwide use of herbicides. However, total herbicide exports accounted for only about 34 percent of the volume and about 50 percent of the value of insecticide exports in the 2 years in question. Fungicides and other pesticides account for a much smaller share of exports than insecticides and herbicides.

In developing countries, virtually all pesticides have to be purchased from abroad at the expense of limited foreign exchange credits. Furthermore, the cost of other inputs such as labor and land values are low compared to the level in developed countries. Under these conditions, the cost of the pesticide input, relative to other inputs, may be as high as 40 percent compared with about 5 to 10 percent in developed countries.

Current Pesticide Shortage

Reports received from around the world indicate that the current and future shortage of pesticides has a real

TABLE 5-4 U.S. Pesticide Exports by Product, 1970-1971

Product	Amount (thousands of pounds) 1970	Amount (thousands of pounds) 1971	Value (thousands of dollars) 1970	Value (thousands of dollars) 1971
Insecticides				
Technical				
DDT	15,299	17,395	2,433	3,187
Aldrin-toxaphene group	22,606	30,459	12,633	16,375
Other poly-chlorinates	17,613	9,770	4,129	2,789
Parathion	10,668	19,018	4,920	7,951
Other organic phosphates	28,470	21,271	22,508	23,129
Other				
Organic	27,257	40,366	20,660	34,692
Inorganic	3,602	2,493	1,078	944
Subtotal technical	125,516	150,772	68,361	89,067
Formulated				
DDT 75% and above	51,758	24,584	11,412	6,284
DDT below 75%	2,493	3,155	512	572
Other poly-chlorinated	7,361	7,938	3,838	3,931
Organic phosphates	13,520	20,845	11,748	17,231
Household	12,627	11,494	9,216	7,737
Other	23,203	18,715	13,945	11,844
Subtotal formulations	110,962	86,731	50,680	45,598
Total insecticides	236,478	237,503	119,041	136,665
Herbicides				
Technical				
Phenoxy	9,571	10,811	3,305	3,802
Other				
Organic	34,586	33,875	29,615	25,042
Inorganic	1,787	1,376	495	448
Subtotal technical	45,944	46,062	33,415	29,292
Formulated	32,017	38,448	28,725	36,627
Total herbicides	77,961	84,509	62,140	65,919

TABLE 5-4 (continued)

Product	Amount (thousands of pounds) 1970	1971	Value (thousands of dollars) 1970	1971
Fungicides				
Technical				
Organic	5,009	6,672	4,336	5,284
Inorganic	4,969	5,629	1,543	2,078
Subtotal technical	9,705	12,301	5,879	7,362
Formulated				
Captan, Hg	1,564	1,320	1,159	759
Dithiocarbamates	12,003	8,916	4,442	3,854
Others	17,761	20,469	12,258	20,549
Subtotal formulations	21,328	30,704	17,859	25,162
Total fungicides	41,306	43,005	23,738	32,524
Other pesticides				
Technical				
Rodenticides	181	220	127	107
Fumigants	33,894	43,120	5,514	6,595
Disinfectants, etc.	5,023	5,033	4,825	4,294
Subtotal technical	39,098	48,372	10,466	10,996
Formulated				
Disinfectants, etc.	5,159	5,539	2,545	2,505
Dips, PGR's	4,157	3,593	2,384	2,992
Subtotal formulations	9,316	9,131	4,929	5,497
Total other pesticides	48,414	57,503	15,395	16,493
All pesticides	404,158	422,521	220,313	251,602

SOURCE: USDA, ASCS; AID.

potential for disaster on the food and health fronts. In addition to the millions of acres of land returned to cultivation in the United States, there are other reasons for alarm about the impending pesticide shortage. The building blocks for pesticides are petroleum derivatives, which have become critically scarce and much more costly. These include intermediates such as ethylene, butylene, propylene, benzene, xylene, phenol, chlorine, and other related materials. Pesticide containers, solvents, and emulsifiers are also petrochemically based. These shortages will affect all countries, whether they are exporters or importers.

Development of Resistance to Pesticides in Arthropods of Medical Importance

The development of resistance to pesticides in pest populations hampers efforts to hold arthropods of medical importance at acceptable population levels. In fact, the development of resistance is a major factor in the increase of malaria in several parts of the world (Wright et al. 1972). While in agriculture the initial impact is the increased dosages (and consequent increased cost) required to achieve satisfactory control (Georghiou 1972a), in malaria eradication and other antivector operations the initial result is a resumption of disease transmission (Busvine and Pal 1969).

Three kinds of resistance have assumed primary importance in public health programs: resistance to DDT, to cyclodiene derivatives (such as dieldrin and gamma-HCH), and to the organophosphorous compounds (OPC) (Brown and Pal 1971). WHO has documented the number of species of public health importance showing these types of resistance as follows:

Year	DDT	Dieldrin	OPC	Total
1956	27	25	1	33
1962	47	65	8	81
1969	55	84	17	102
1974	61	92	27	109

With the use of propoxur, a relatively new residual insecticide for malaria control, resistance to carbamate insecticides has become a matter of concern as well. Of anopheline mosquitoes, 38 species have developed populations resistant to some insecticide, 36 to dieldrin, 15 to DDT, and 1 to malathion and propoxur (WHO 1971). Nevertheless, there still remain 31 species of proven malaria vectors that have not yet been found to have developed resistance (Pal 1974).

The problem of resistance to DDT in insects of health importance was recognized as early as 1947, when *Musca domestica* and *Culex fatigans molestus* were reported resistant to DDT in Italy. Anopheline resistance was reported in Greece and the United States in 1949. WHO began an intensive program of research on the problem in 1956, when two malaria vectors, in Indonesia and Nigeria, developed resistance to DDT and dieldrin, respectively. Over 10 years ago, the deputy director of the Malaria Institute of India made this statement: "It is now generally accepted that resistance to insecticides by all disease vectors is inevitable" (Pal 1962).

This statement has, perhaps, a more fatalistic ring than more recent considerations, because, as Brown and Pal (1971) point out in their revised review of the subject, "there is some room for strategy in the use of existing insecticides." For instance, one insecticide may be used for a series of generations, followed by a switch to another for a series of generations. Even if it is not possible to delay the development of resistance long enough for the eradication of the disease parasite, the absence of selection pressure from the chemical does allow for at least partial reversion to a state of susceptibility (unless similar compounds are used agriculturally). In any case, another statement made by Pal (1962) still holds true today--that "one of the most important consequences of the resistance problem" has been "the realization that chemical control cannot be taken as a substitute for sanitation practices, and that these methods are complementary to each other in the control of arthropod borne diseases." Nevertheless, in spite of the utilization of these strategies, the problem of resistance to pesticides will be with us for the foreseeable future.

It is not always clear why resistance develops. In certain cases, however, it is evident that agricultural use of pesticides is the primary cause. Central America provides one of the clearest of such cases. Figure 5-1, from a Pan American Health Organization status report, (PAHO 1972) shows areas of North and Central America that experienced "technical problems" with their malaria eradication programs in 1971. In all areas of Central America shown, except for parts of Mexico (labeled 9 and 10), the technical problems are due at least in part to resistance to DDT, to DDT and dieldrin (in Honduras), or to DDT and malathion (in Nicaragua). The report states that "the

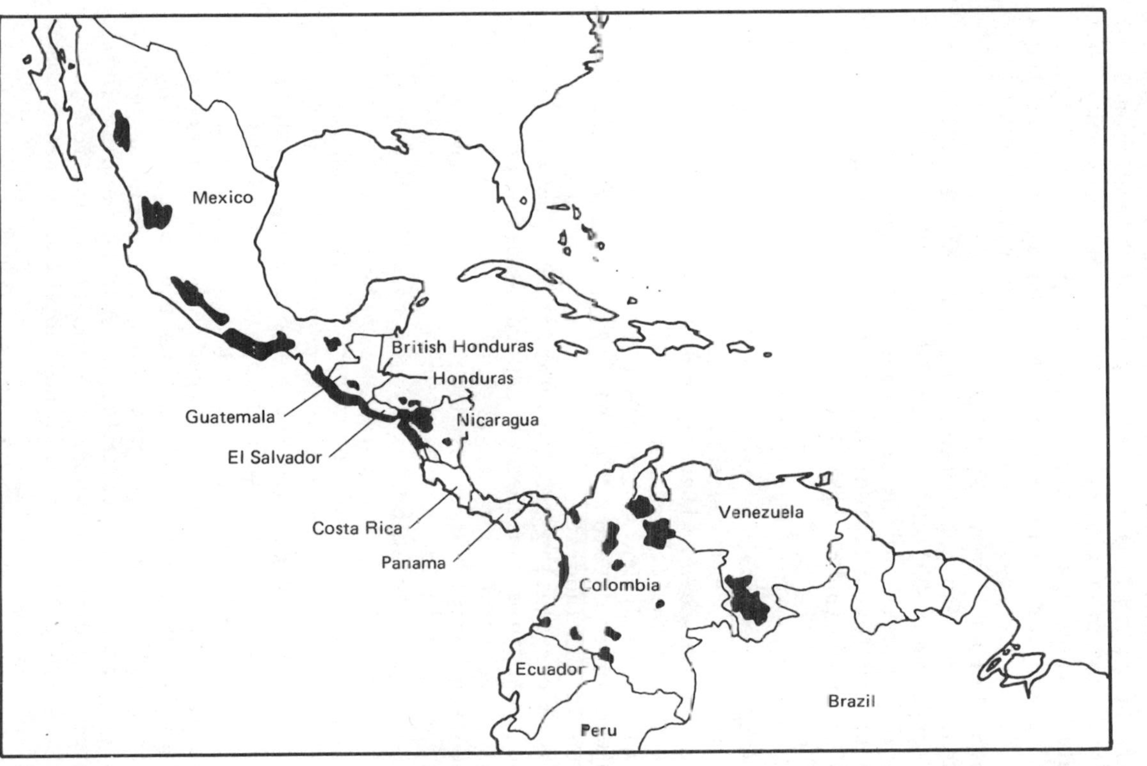

FIGURE 5-1 Geographical distribution of areas with technical problems in malaria eradication programs.

SOURCE: PAHO 1972.

vector's physiological resistance to the commonly used insecticides such as DDT and dieldrin has been verified in El Salvador, Guatemala, Honduras, Mexico and Nicaragua, in the Pacific coast zones where various kinds of insecticides are being applied in cotton-growing." Additional research by national malaria eradication programs shows that "areas where the specific insecticides are widely used in agriculture generally overlap the areas where the malaria vector is resistance to those insecticides" (García Martín and Nájera-Morrondo 1972).

Evidence has been presented (Georghiou et al. 1972, 1973; Georghiou 1972b) that strongly suggests that in cotton-growing areas of Central America a correlation exists between the extent of agricultural pest control operations on cotton and the level of resistance in malaria mosquitoes:

1. *Anopheles albimanus* collected in Haiti developed only a three-fold resistance to organophosphate and carbamate insecticides even with rigorous selection in the laboratory. On the other hand, *A. albimanus* collected in El Salvador already had a threefold to fourfold resistance level and quickly developed levels over thirtyfold with selection in the laboratory.
2. Good correlation exists between the spectrum of resistance found in mosquitoes in Central America and the types of insecticides used in agriculture.
3. Cotton insects have developed resistance in Central America, especially to parathion.
4. Mosquito populations are found to decline during the wet season. If agricultural pesticides were not used during that season, one would expect a rise in population.
5. The geographical range of resistance is complex. It is difficult to correlate the range directly with the amount and frequency of pesticide application, because those records are often not kept. Resistance does seem to correlate with the size of holdings, however. Large holdings generally mean more applications and more spraying done by air rather than from the ground. Other factors are also important. For example, resistance is high in the center of Nicaragua, where the insect population is geographically isolated as well as subjected to heavy treatments of pesticides.
6. Seasonal data show a rise and fall of resistance among mosquitoes paralleling the rate of use of agricultural pesticides. Where the pesticides have been used for a number of years, the resistance level fluctuates less with the seasons.

Problems other than the resistance of malaria mosquitoes have also been aggravated by the use of pesticides in Central America. In Nicaragua, about 80 percent of the population consists of subsistence farmers who grow mainly corn and beans. In some areas the disruption of the agroecosystem of the region by pesticides applied to cotton has been disastrous for these farmers. The species of leafhopper that transmits cornstunt has become a serious pest because of a disruption of natural controls in areas where pesticides have been heavily used. In some of these areas, the peasants can no longer grow corn successfully.

In most of the Central American regions that have experienced problems due to vector resistance, the carbamate propoxur has been substituted at greater cost but with quite good results. However, according to the PAHO status report, "In August 1970 it was found that in the Rio Viejo valley in Nicaragua the local *A. albimanus* was resistant to propoxur. The area of resistance is a valley in which rice is extensively cultivated and which has been subject to aerial spraying with carbaryl for a number of years." Other such foci, also associated with carbaryl use on rice, were found in 1970 and 1971 in El Salvador (PAHO 1972). Obviously, a potentially disastrous situation exists, which is likely to cause even greater setbacks in the Latin American malaria eradication programs unless ministries of health and agriculture can coordinate their activities.

Agricultural pest control activities have interfered with malaria eradication in other parts of the world as well. In Egypt, "the pre-existing double resistance [to both DDT and dieldrin] in *An. pharoensis* produced by agricultural insecticides dictated the use of HCH when the malaria eradication campaign was started in 1966" (Brown and Pal 1971). Agricultural pesticides have also been implicated, in part at least, in anopheline resistance in Greece, Iraq, and Java. An analogous situation was the induction of malathion resistance in lice in Burundi by the withdrawal for personal use of pesticide formulations originally destined for coffee pest control.

It is interesting that few cases of resistance due to agricultural spraying have occurred in Asia. Resistance has arisen in India, in areas where wells were treated with larvicides and where intensive spraying against malaria has gone on for a number of years. Some parts of Indonesia and a few other Asian foci have had problems with resistance. Actually, agricultural spraying has been relatively light in Asia thus far. Coordination between health and agricultural planners in this part of the world would probably be most valuable at this time.

Hamon (1970) has discussed the development of resistance in parts of Africa. The situation there is slightly different, because in much of Africa little or no spraying has yet been done for malaria control, despite the prevalence of the disease and the vector. Hamon states:

> In some parts of the world, such as West Africa, we have had very little selection of resistant *Anopheles* populations by health department application of insecticide. Most, if not all, of our resistant populations came from BHC dusting of cocoa and coffee in the south, DDT and endrin dusting of cotton in the middle belt, and DDT-BHC dusting of the peanuts in the north . . . it is probable that at least in Africa, where malaria-control operations have been very much delayed, we shall select resistant populations of the vector before any application of the insecticides by the health departments.

Although the WHO-initiated malaria eradication project has made very great strides, still there are a number of areas in the world where eradication has not even been attempted and malaria is as widespread as ever. Programs do not exist in most African countries south of the Sahara, in most of the Amazon Valley region, in parts of Indochina, or some islands of Southeast Asia and Oceania (WHO 1970). It would be a great mistake to undermine the possibility of such programs through extensive agricultural spraying, just as it was a mistake to allow spraying of cotton in Central America to such an extent that ongoing malaria control programs were jeopardized. Yet, despite the arguments of Hamon, García Martín, and others, only limited effort has been made to coordinate plans for areas where arthropods of agricultural and public health importance are under simultaneous attack by insecticides (MacDonald 1972).

As the development of resistance to DDT has reached levels that make the operational use of DDT unfeasible in antimalaria programs, it becomes necessary to consider replacement insecticides and their costs. The majority of the countries operating malaria eradication programs at present have difficulty in meeting the financial obligations even when DDT, an extremely cheap insecticide, is used. All new insecticides available as alternatives are more expensive than DDT, and their adoption on a wide scale would certainly create serious financial problems. It has been estimated that 35,000 metric tons of DDT are currently being used each year in antimalaria programs.

The annual cost of spraying operations with DDT is estimated to be $60 million. The annual cost, should it be necessary to substitute malathion or propoxur, at present would be well over the 1972 estimates of $180 million and $500 million (Wright et al. 1972). Of nearly 1,500 chemicals that WHO has tested in the field, only two--malathion and propoxur--have been approved as satisfactory substitutes for DDT in antimalaria programs. Two others--fenitrothion and methyl dursban--have achieved conditional approval. With so few pesticides available to control malaria vectors, the development of resistance to any one of them is a matter for concern.

Resistance in pest populations can be expected to develop to chemicals other than the traditional insecticides. Resistance has been induced by laboratory selection of *Aedes aegypti* to the chemosterilants apholate and metepa. Resistance to new control agents such as juvenile hormone mimics and pheromones can also be anticipated, since it has been induced to methopurene by laboratory selection of *Culex*.

One common suggestion to alleviate resistance problems is that certain chemicals be reserved exclusively for public health uses. Many experienced investigators of insect resistance do not believe this will be practical for several reasons. First, no one pesticide could be depended upon to take care of all the different species of public health pests. Second, no compound is known that is different enough from the others to avoid the problem of cross-resistance. Even the juvenile hormone analogues have suffered from some degree of cross-tolerance in houseflies resistant to the conventional insecticides.

Use of Insecticides in Cotton Production in Central America

The fertility and condition of the soil, the available moisture, ambient temperatures, cotton varieties, and human resources required for successful commercial cotton production are present over much of the Pacific plain of Central America. Above all else, however, cotton pests are the principle factors that determine the success or failure of cotton production in the entire area. While cotton pests are a threat wherever cotton is grown in the world, it is unlikely that nay other region has more serious pest problems than the cotton-producing areas of Central America. As an example of the sequence of events and the resulting complications, the story of Nicaragua will be presented, as it is the best documented of the region (Falcon and Daxl 1973).

There are at least ten genera of major insect pests that interfere with cotton production in Nicaragua. While most of these genera are represented by only one species, the genus *Spodoptera* is represented by at least six species. In addition to insect pests, there are several plant diseases, of which boll rots are the most important.

The activity and abundance of the cotton pests are regulated by many factors, including rainfall, temperatures, the state of crop development, predation by natural enemies, and diseases. The boll weevil is the key pest--it is a highly injurious, perennially occurring, persistent species that dominates control practices. In the absence of deliberate control efforts by man, the boll weevil populations often exceed the economic injury level. The boll weevil is considered to be indigenous to the area and thrives on commercial, wild, and volunteer cotton plants as well as other plants in the family Malvaceae. While there are natural enemies of the boll weevil, its abundance is regulated mainly by the condition of the host plants. This pest is the major obstacle to the development

TABLE 5-5 Changes in Importance of Cotton Insect Pests in Nicaragua

Species	Common Name	Order of Importance[a]		
		1956[b]	1965[c]	1970[d]
Alabama argillacea	Cotton leafworm	2	5	4
Anthonomus grandis	Boll weevil	1	3	1
Aphis gossypii	Cotton aphid	4	M	M
Bemisia tabaci	Whitefly	*	4	3
Creontidades spp.	Plant bugs	*	M	6
Heliothis spp.	Bollworm	3	1	2
Prodenia spp.	Armyworm	*	2	5
Sacadodes pyralis	False pink bollworm	5	*	*
Spodoptera spp.	Armyworms	*	7	8
Trichoplusia ni	Cabbage looper	*	6	7

[a]Number 1 denotes most important; *, not important, and M, of minor importance.
[b]Source: OIRSA, 1956 Lista de enfermedades de los cultivos principales existentes.
[c]Source: Smith 1972.
[d]Source: Falcon 1971.

and establishment of integrated control programs on cotton in Central America. Other important cotton pests in Nicaragua are given in Table 5-5. Table 5-6 summarizes the importation and use of insecticides in Nicaragua from 1965 to 1971. Table 5-7 shows the annual cotton production from 1962 to 1973 and the value of exports.

Of the plant diseases, the boll rots present the most serious problems and cause more damage than most other pests. Not only does their damage contribute significantly to increased production costs, but they destroy up to 25 percent of the harvestable crop.

Commercial cotton production in Nicaragua has relied almost totally on the use of chemical pesticides for pest control, and without chemical pesticides the cotton industry could not have developed, nor can it continue to survive. During the 7-year period of 1965-1971, an average of $9.7 million (U.S.) worth of chemical pesticides were imported yearly into Nicaragua, with over 70 percent of this material used on cotton. Revenues generated by the cotton industry have added significantly to the gross national product, provided employment, and contributed to population growth.

On the negative side, cotton production techniques have led to the development of many problems that have had deleterious economic, social and environmental effects. Some of the most important problems are

1. Human intoxications by chemical pesticides, which yearly result in the sickness and disability of thousands of farm workers (Adam 1972).
2. Contamination of human milk, which, as shown by data gathered in the cotton-growing areas of Guatemala, results in suckling children's consuming from 6 to 207 times the maximum daily acceptable intake of DDT advocated by FAO/WHO (Olszyna-Marzys et al. 1973).
3. Contamination of meat sold for human consumption.
4. High levels of pesticide residues in estuarine and marine fish and invertebrates as demonstrated by a study conducted in estuarine waters in the cotton-growing areas of Guatemala (Amado de Zeissig 1973).
5. Creation of new pest problems on cotton (Falcon 1971).
6. Aggravation of pest problems on surrounding crops such as vegetables and corn (corn-stunt disease is a serious problem affecting corn grown in the cotton-producing areas of Nicaragua) (Tapia 1972).
7. High levels of resistance to chemical insecticides exhibited by cotton pests such as *Heliothis zea* (Wolfenbarger

TABLE 5-6 Importation and Use of Insecticides in Nicaragua 1965-1971

Year	Value (in millions of U.S. dollars) of Insecticides and Adjuvants Imported[a]	Percent Used in Cotton[b]	Value (in millions of U.S. dollars) of Insecticides Used in Cotton	Value (in millions of U.S. dollars) of Insecticides Used in Cotton per Manzana[c]
1965	10.5	86.8	9.1	40.80
1966	11.0	86.6	9.5	44.16
1967	13.1	84.4	11.1	52.90
1968	10.6	79.6	8.4	44.96
1969	6.8	75.1	5.1	32.72
1970	7.0	72.0	5.0	36.74
1971	10.4	72.1	7.5	48.17

[a] Source: Ministerio de Agricultura y Ganaderia, Nicaragua.
[b] Source: Banco Central de Nicaragua.
[c] One manzana = 0.7 hectare.

TABLE 5-7 Cotton Production and Value of Exports in Nicaragua 1962-1973

Year	Area Planted (ha × 1,000)	Average Yield Seed Cotton (kg/ha)	Approximate Value (in millions of U.S. dollars) of Cotton Lint and Seed Exported
1962	93.9	2,270.8	45
1963	115.4	2,364.3	58
1964	133.9	2,663.0	75
1965	142.0	2,264.3	65
1966	150.7	2,228.6	62
1967	146.4	2,066.2	46
1968	131.4	2,066.2	46
1969	108.6	1,840.3	50
1970	95.4	2,481.8	50
1971	109.3	2,701.9	70
1972	161.0	1,785.7	60
1973	200.2	2,116.9	70

SOURCE: Comision Nacional del Agodon, Nicaragua.

et al. 1971) and by nontarget insects that occur in the vicinity of cotton fields, such as the malaria vector mosquito, *Anopheles albimanus* (Georghiou 1972b).

Because of the difficulties encountered in controlling *A. albimanus*, the incidence of malaria increased greatly during the 1960s and was reversed only with the introduction of the carbamate insecticide propoxur (OMS 33) for control of *A. albimanus*. However, the cost of using propoxur for *A. albimanus* control is $11 to $12 (U.S.) per house per annum compared to $1 to $2 for DDT. Thus, with a given yearly allotment, the Nicaraguan malaria control program could protect almost six times as many homes with DDT as it could with propoxur. The problem was reduced temporarily in 1973 because Bayer Industries of West Germany donated 98,000 pounds of propoxur (Baygon) to Nicaragua.

Along with malaria, most of the human inhabitants of the cotton-growing areas of Nicaragua are exposed to other arthropod-borne diseases such as Chagas' disease, dengue fever, and equine viral encephalitis. In addition to disease, these people suffer from malnutrition. One contributing factor to the problem of malnutrition is that farmers have great difficulty producing beans, maize, and other crops at a subsistence level because these crops suffer so heavily from pest attacks. Pests of corn and beans occur at outbreak levels largely because of destruction of natural enemies by the application of pesticides to cotton. As discussed by Scrimshaw et al. (1968) synergism between malnutrition and infection is responsible for much of the excess mortality among infants and preschool children in less-developed regions of the world. This condition was shown to be a serious problem in Guatemala under conditions similar to those that exist in the cotton-growing areas of Nicaragua.

Thus, while chemical pesticides are primarily responsible for the phenomenal growth and development of the cotton industry in Nicaragua, they have also contributed significantly to human misery and death among the farm workers, recurrent crop failures in corn and beans, greatly increased costs of cotton production, economic instability, and serious destruction of the environment. It has been estimated (Falcon and Daxl 1973) that while the release of about $10 million (U.S.) worth of chemical pesticides into the cotton agroecosystem every year provides a return of $60 million or more per year in revenues derived from cotton exports, the yearly cost to society and the environment for this exchange is about $200 million when reasonable costs are assigned to the economic, social, and environmental problems brought on by the use of pesticides.

PUBLIC HEALTH PEST CONTROL INCIDENTAL TO AGRICULTURAL PESTICIDE USE

The preceding sections paint a grim picture of the effects of the use of agricultural pesticides on the public health. Although these negative effects have been very significant, it is undoubtedly also true that much of the pesticide usage in agricultural practice has unanticipated beneficial effects in its incidental impact on insects of public health importance, particularly the mosquitoes that breed in rice fields. A well-documented example involves the use of granular formulations of fensulfothion or carbofuran applied to control the root-feeding larvae of the rice water weevil in Arkansas; the chemicals gave complete control of mosquitoes with the application of 0.5 lb/acre just before the fields were flooded (Lancaster and Tugwell 1969). Similar results were obtained in Louisiana with granules of carbofuran and the carbamate Bux applied from aircraft within 5 days after flooding. The use of seed treated with these insecticides for the control of the rice water weevil has been found to result in partial control of *Psorophora* mosquito larvae.

During the 1973 outbreak of the fall armyworm on rice in Louisiana, it was found that the extensive preflood applications of toxaphene at 2 lb/acre to meet the emergency resulted in complete elimination of *Psorophora* larvae, as well as of other aquatic organisms. In that state it is found that early-season seeding ("water-planting") from aircraft results in much lower mosquito production than drill-planting of the rice seed later in the season. The draining and reflooding involved in the routine of applying fertilizers is a great mosquito producer in Louisiana as in other states. Rice fields treated with herbicides offer greater opportunities for larvivorous fish such as *Gambusia* to reach their mosquito prey unhindered by decumbent vegetation (Craven and Steelman 1968).

The best-documented case of agricultural insecticides reducing numbers of insect vectors of disease concerns Japanese encephalitis (JE) in the Republic of Korea, where in 1966 there were 3,600 reported cases, of which 1,100 were in an epidemic in Cholla Pukdo province. Since 1969 the number of cases annually has not exceeded 300 in South Korea. This reduction has been correlated with the great reduction in the vector *Culex tritaeniorhynchus*, confirmed by light-trap surveillance figures during the period. It has now become quite difficult to find larvae of this mosquito in rice fields, although they persist in swampy land and in vacant lots around the growing towns and cities (Self et al. 1971).

During this period there has been a great increase in the volume of insecticides applied against rice pests, principally the stem borer and two species of leafhopper. The volume in terms of cash cost increased from $6.1 million in 1966 to $15.4 million in 1969; the weight of insecticides imported increased from 297 tons in 1968 to 849 tons in 1970 (Bang and Self 1971). The volume continues to increase as heavy import duties are reduced. The rice insecticides, which are mainly fenitrothion, EPN, phenthoate, and trichlorfon, are typically applied in two treatments, the first at 0.3 kg/ha, and the second at 0.9 kg/ha; both rates have been found by test to eliminate the larvae of *Culex tritaeniorhynchus* and prevent reappearance of larvae in the fields for about a week. Accordingly, the recent encephalitis cases in Korea have been mainly on the outskirts of cities where farms have not been kept up, and particularly in the vicinity of swamps or river floodplains.

The danger is that the vector mosquito will develop resistance to the organophosphorus insecticides employed. This has not yet happened in Korea; in Taiwan, however, where the JE cases are sporadic but more numerous than in Korea and there are three species of vectors (*C. tritaeniorhynchus*, *C. annulus*, and *C. fuscocephalus*), there are indications of resistance to temephos in *C. fuscocephalus* in the southeast part of the island. In northern Taiwan, the ULV airsprays of malathion at 0.65 lb/acre applied against the green rice leafhopper were found also to give short-term reduction of 50-80 percent in the adult population of *C. annulus* (Mitchell et al. 1972).

AN AGROMEDICAL TEAM APPROACH TO PESTICIDE MANAGEMENT

With present population trends, increased food and fiber production is one of the central goals for all nations. In the past, agricultural chemicals have been the mainstay of all expanding food production programs and their continued use will be essential to meet these increased demands. The growing problems of pest resistance to pesticides and of environmental degradation, however, indicate that the future use of pesticides will have to be part of a multifactorial pest control program utilizing a wide variety of pest management techniques, of which the use of chemicals will be but one component. No longer can reliance be placed solely on chemical control. The future use of pesticides will have to be more judicious and restrained, respecting the benefits of natural enemies in regulating harmful pests.

Pesticide usages and pest management practices have undergone profound and rapid changes in the last decade. Problems of pest resistance, primary and secondary pest resurgences, increased production and registration costs for pesticides, increased human toxicity problems, environmental degradation, and changing agricultural and social patterns are some of the factors that are responsible for these changes. Pest management surveys conducted in widely different geographic areas by multidisciplinary teams for the University of California/AID Pest Management Project have reported that in many areas of the world the use of agricultural pesticides is poorly understood and in certain areas it is imprudent and excessive (Apple and Smith 1972; Cavin et al. 1972, Echandi et al. 1972, Sasser et al. 1972, Glass et al. 1972).

Serious problems, some of them with far-reaching consequences, have been encountered and have resulted, on occasion, in appeals to international organizations for assistance and special training in pesticide management. Assistance has been sought also for integrated pest control guidance and advice on pesticide legislation and registration. The areas of most concern have been residue contamination of food with pesticide levels too high for international export, increases of human pesticide intoxication, and malarial upsurge due to vector resistance acquired as a result of agricultural pesticide use.

During a recent visit by project representatives of the University of California/AID Pest Management Project to El Salvador, examples of all these problems were encountered.

Pesticide Management Problems in El Salvador

El Salvador is on the Pacific coastal plain of Central America and covers an area of 21,000 square kilometers; the population is approximately 3.8 million. Much of the country is hilly or mountainous, and cotton is the major agricultural commodity; this crop is largely confined to the valleys and coastal plains. Substantial amounts of pesticides are used for agricultural production and in puboic health for malarial control. It is estimated that

8,000,000 pounds of pesticides are imported annually, 60 percent of which are ethyl and methyl parathion. Most of this is used on cotton, often with 18 to 20 applications to a crop. A mixture of DDT and toxaphene is also used. There have been numerous reports of pesticide poisonings. Although human poisoning is not a notifiable disease, excellent records of hospitalized cases of pesticide poisonings from 1963 to 1972 were kept by the public health authorities in the Ministry of Health (Valverde and Ruegas, Personal communication, 1973):

Year	Nonfatal Cases	Fatal cases
1963	1,104	11
1964	965	2
1965	938	1
1969	584	7
1970	474	7
1971	586	10
1972	2,787	5

The data for 1972 are important not only because they identify a greater than fourfold increase over the preceding year, but also because they were confirmed by another survey conducted by representatives of the Ministry of Labor who surveyed all areas of the country, obtaining hospital and nonhospital incidence data of pesticide poisoning. Their survey identified 2,028 pesticide poisoning cases, including 202 women and 224 children; there were 25 adult male fatalities, 2 adult female fatalities, and 3 child fatalities (more adult males were exposed during application). The authorities felt that the three major factors contributing to the increase in that year were (a) the greatly expanded use of ultra-low-volume applications of ethyl-methyl parathion mixture on cotton; (b) the original selection of a 72-hour re-entry time to fields by workers after applications, which is now considered to be too short an interval; and (c) an unusually dry season with little rainfall to wash off the pesticide residue. This pesticide intoxication problem is one of special concern to public health and occupational health authorities.

Pesticide Poisonings in Other Areas

Surveys in other parts of Central America have indicated that human pesticide poisoning is a serious problem in areas other than El Salvador. Where cotton is the major crop, ethyl and methyl parathion mixtures have been widely used. There are several statistical reports on organophosphate poisonings. In a report from WHO, Vandekar and coworkers (1971) published the following poisoning statistics from Guatemala and Nicaragua:

Year		Guatemala	Nicaragua
1968		1,374	
1969		837	258
1970		659	221
1971		1,100	356
	TOTAL	3,970	835

From a global point of view, reports of widespread poisoning in pesticide workers and in the general population are numerous. Although accurate statistics are not available, it is obvious that the problem is both extensive and serious. The World Health Organization estimates that there are approximately 500,000 cases occurring annually, with a mortality rate of more than 1 percent (WHO 1973).

Pesticide poisonings were also recognized in Indonesia in 1960 when the organochlorines were the major offenders. Even incomplete data acquired in Java and northern Sumatra revealed that 478 organochlorine poisonings and 31 organophosphate poisonings had occurred, and there were reports of 125 deaths (Davies 1972). Most of the data listed only poisoning incidents that involved more than one person. The most common mechanism for poisonings was the accidental contamination of food from discarded pesticide containers. There were numerous reports of groups of people who became ill, vomited, and had convulsive seizures following ingestion of food prepared in a pesticide container. The diagnosis was usually made on epidemiological grounds and was seldom confirmed by analytical toxicological studies.

The inherent toxicity of the kinds of pesticides used appears to be one of the most important variables in determining extent of poisoning problems. Thus, as more and more countries switch from the organochlorine pesticides to the generally more toxic organophosphate pesticides because of increased pest resistance, more and more reports of human intoxication will occur unless the introduction of the chemicals is preceded by a virorous pesticide management training program.

When the organophosphates were first widely used in the United States, similar problems occurred. Itinerant salesmen peddled parathion in unlabeled brown bags and sold the product as a "roachkiller"; parathion could be bought in the local stores and even the local barbershops, and headlines telling of pesticide poisonings were common in the local papers. Cases occurred in adults and children alike and were especially frequent among the black population (Davies et al. 1965 1970; Reich et al. 1968). Parathion was frequently brought into the urban setting from the fields via discarded pesticide containers that often contained small amounts of pesticide residue (Davies et al. 1973).

The inherent toxicity of parathion was poorly understood, and cases occurred as a result of accidents, worker exposure, and attempts at suicide and homicide. It was not until appropriate legislation was introduced that the situation began to improve. Publicizing the dangers of this agricultural chemical in the mass media was another important factor in increasing awareness of the danger of pesticides. During 1965-1972 the situation improved, as can be seen from the statistics from Dade County and from the State of Florida (Table 5-8). Clinical recognition and management of this medical emergency has also improved, contributing to the declining mortality rate. Organophosphate poisoning can mimic epilepsy, pneumonia, diabetes, and many other conditions. Consequently, in the early days, intoxication was not readily identified. Early diagnosis

TABLE 5-8 Parathion Poisoning Cases in Dade County, Florida, and Pesticide Mortality and Morbidity in the State of Florida, 1965-1972

	Dade County		Florida	
Year	Fatalities	Nonfatalities	Mortality	Morbidity
1965	5	14	22	649
1966	6	7	22	n.a.
1967	4	6	35	437
1968	2	5	26	484
1969	3	3	17	365
1970	2	7	13	299
1971	0	5	8	323
1972	0	5	7[a]	n.a.
1973	1	8	n.a.	n.a.

[a]Incomplete data.

SOURCE: Bureau of Vital Statistics, Department of Health and Rehabilitative Services of Florida State, 1973.

is essential for successful treatment of this type of poisoning. The case fatality rate for parathion poisoning in Dade County, a rate that expresses the chances of dying from a particular condition and that is in part a measure of the efficacy of medical treatment, dropped from 18 percent in 1965 to 0 in 1972 (Table 5-8). These data confirm the improved clinical management of this condition over the years. Similar improvements can be expected in those areas of the world now experiencing epidemic pesticide poisoning. Legislation, education, and training of workers will be the key to the problem.

Persistence

Persistence and the consequences of food and human contamination was a second problem of pesticide management that was readily appreciated in El Salvador. In addition to the cotton production, a sizable meat industry has developed. The main grazing pasture is on the coastal plains in areas adjoining the cotton fields. In addition to ethyl and methyl parathion, a DDT-toxaphene mixture was being used on cotton to control pests. Aldrin, the main pesticide used to control corn pests, was used as a preplant treatment and also for treatment of the crop after emergence. There was, therefore, ample opportunity for contamination of cattle. Reports of excessive pesticide residues in beef cattle were numerous, and on several occasions exportation of beef lots to the United States has been prohibited on account of excessive pesticide residues. These restrictions posed a serious threat to the meat export industry and the general economy of the country.

Although these contaminations were of serious concern to cattlemen and to agricultural officials, concern about the residue problem was also expressed by representatives of the Ministry of Health. Health officials were concerned that food designated for export, but that failed to meet international standards, might be redirected for domestic human consumption and thus endanger the people of the country. They regretted the absence of baseline information on levels of persistent pesticides among the general population and expressed the hope that future residue-monitoring programs of food and feed stock would also include fat analyses obtained from persons undergoing surgery.

Resistance

Resistance was the third pesticide management problem encountered in El Salvador. Malaria was incompletely controlled and Dr. Oscar Nave-Rebollo reported at the 9th Annual Public Health Congress that "the malarial situation of the country at the end of 1970 had not improved substantially and in fact the situation had deteriorated" (Nave-Rebollo 1971). Malariologists were concerned about the appearance of mosquito resistance to the carbamate insecticide propoxur as well as to DDT. The greatest incidence of malaria in El Salvador is along the coastal plains. *Anopheles albimanus*, the main malaria vector in El Salvador, is resistant to DDT, dieldrin, and propoxur in some areas. As has been reported in other areas of the world and discussed earlier here, agricultural pesticide usage has interfered with control of malaria.

An Agromedical Seminar

The recognition of these pesticide management problems prompted the organization of a pesticide management training program in El Salvador in December 1973. The program was developed by the University of California/AID Pest Management Project with assistance from the Pan American Health Organization. The emphasis was on an agromedical approach, with in-country representatives formulating their own recommendations for solutions to the problems. The theme of the seminar was "Pesticide Management in El Salvador," with emphasis on the problems of persistence, resistance, and poisoning.

Through a multidisciplinary agromedical approach, methods of pesticide management as a preliminary step to the ultimate goal of integrated pest control were developed. The topics ranged from chemistry to toxicology and the chemodynamics of pesticides, development of resistance, pesticide contamination, epidemiological aspects of pesticides, integrated pest control, and malaria, as they relate to El Salvador.

In order to obtain active involvement of the people of the country, a technical steering committee was selected and each member joined one of the working groups. The working groups identified the specific pesticide management problems of their country and then sought to enunciate objectives and devise programs and methods of implementation of programs for the solution of those problems. Following the seminar, an intensive training program for chemists was conducted for 2 weeks. In this program, time was spent in the laboratory, reviewing instrumentation and introducing new analytical methodology for pesticide analyses.

Evaluation of any training seminar is not easy, and with a topic as complex as pest management, progress can be measured only some time after the initial training endeavor. This is certainly the case for the ultimate evaluation of the program in El Salvador. Certain immediate responses observed in the training program, however, suggest that the initial impact of this program was highly successful and augur well for the future.

The results of the workshop can be measured in part by the recommendations and resolutions formed by each technical group. These were truly reflective of the multidisciplinary agromedical approach. The subject matter of the proposals ranged over a wide variety of topics. The organization in the near future of a central interagency pesticide commission was proposed, to include representatives from the Ministries of Agriculture, Public Health, Labor, Economy, and Defense and also from the National University of El Salvador and private industry. This commission, which would be inaugurated by Executive Decree, would legislate and coordinate the use of pesticides for the development of the country and the protection of the population and the environment. It would develop pesticide residue tolerances, develop legislation to control the importation of new pesticides, and thoroughly investigate pesticide protection plans.

Joining and complying with the requirements of the Codex Alimentarius Commission of the UN was recommended, a step that would require continued upgrading of laboratory performance and the establishment of a quality-control program. Collaboration in activities related to pesticide residue control in Panama and other Central American countries, and participation in the work of regional bodies such as INCAP and ICAITA was stressed. The toxicological recommendations included the establishment of a chair of toxicology in the medical school and the training of members of regional multidisciplinary pesticide protection teams. It was recommended that the latter should be capable of investigating, analyzing, and reporting human and environmental pesticide poisoning problems to the central commission. Ongoing blood and adipose surveys in man were recommended, along with studies of crops and livestock.

The recommendations for training and education proposed the future education of chemists, public health personnel, and agronomists who would be supported by an updated pesticide reference library.

The breadth of these recommendations, developed and approved by agronomists, public health people and industry, are a measure of the success of the conference. All participants agreed that these recommendations would be presented to the appropriate ministers for their final consideration.

REFERENCES

Adam, A. V. (1972) Summary of Joint FAO/Industry Seminar on the Safe, Effective and Efficient Utilization of Pesticides in Agriculture and Public Health in Central America and the Caribbean (TF:LAT/16). Food and Agriculture Organization of the United Nations. AGPP: MISC/6.

Amado de Zeissig, J. A. (1973) Investigación de Insecticidas Residuales en la Fauna Marina de Los Esteros de la Costa Sur de Guatemala. Universidad de San Carlos de Guatemala.

Apple, J. L. (1972) Intensified pest management needs of developing nations. BioScience 22(8):461-464.

Apple, J. L., and R. F. Smith (1972) A preliminary study of crop protection problems in selected Latin American countries. UC/AID Pest Management Project Rep.

Bang, Y. H., and L. S. Self (1971) Pesticide spray practices in rice fields and control of culicine mosquitoes in South Korea. Unpublished document. WHO/VBC/71.269.

Brown, A. W. A. (1972) The ecological implications of insecticide usage in malaria programs. Am. J. Trop. Med. Hyg. 21(5):833-834.

Brown, A. W. A., and R. Pal (1971) Insecticide Resistance in Arthropods. Geneva: World Health Organization. p. 10.

Busvine, J. R., and R. Pal (1969) The impact of insecticide-resistance on control of vectors and vector-borne diseases. Bull. WHO 40:73-74.

Castillo, T. (1970) Impact of agricultural innovation on patterns of rural life (Focus on the Philippines), pp. 13-52. *In* Agricultural Revolution in Southeast Asia, Vol. II. Consequences of Development. Report of the Second SEDAT International Conference on Development in Southeast Asia, New York, June 24-26, 1969.

Cavin, G. E., D. Raski, R. G. Grogan, and O. C. Burnside (1972) Crop Protection in the Mediterranean Basin; A Multi-Disciplinary Team Report. UC/AID Pest Management Project Rep.

Craven, B. R., and C. D. Steelman (1968) Studies on a biological and a chemical method of controlling the dark rice field mosquito in Louisiana. J. Econ. Entomol. 61:1333-1336.

Davies, J. E., J. O. Welke, and J. L. Radomski (1965) Epidemiological aspects of the use of pesticides in the south. J. Occup. Med. 7:12.

Davies, J. E., J. S. Jewett, J. O. Welke, A. Barquet, and J. J. Freal (1970) Epidemiology and chemical diagnosis of organophosphate poisoning. *In* Pesticide Symposia, W. B. Deichman (ed.). Miami: Halos and Associates.

Davies, J. E. (1972) Recognition and management of pesticide toxicity. *In* Advances in Internal Medicine, G. H. Stollerman (ed.). Chicago: Yearbook Medical Publishers.

Davies, J. E., J. C. Cassady, and A. Raffonelli (1973) The pesticide problems of the agricultural worker. *In* Pesticides and the Environment: A Continuing Controversy, W. B. Deichman (ed.). Miami: Symposia Specialists.

Echandi, E., J. K. Knoke, E. L. Nigh, Jr., M. Shenk, and G. T. Weekman (1972) Crop Protection in Brazil, Uruguay, Bolivia, Ecuador, and Dominican Republic. UC/AID Pest Management Project Rep.

Falcon, L. A. (1971) Progreso del control integrado en el algodón de Nicaragua. Rev. Peru. Entomol. 14(2):376-378.

Falcon, L. A., and R. Daxl (1973) Report to the Government of Nicaragua on the Integrated Control of Cotton Pests, 1970-73. Rome: FAO.

Farvar, M. T., and J. P. Milton (1973) The Careless Technology, Ecology and International Development. London: Tom Stacey.

Gárcia Martín, G., and J. A. Nájera-Morrondo (1973) The interrelationships of malaria, agriculture, and the use of pesticides in malaria control. Ofic. Sanit. Panam. VI(3):20.

Georghiou, G. P. (1972a) The evolution of resistance to pesticides. Annu. Rev. Ecol. 3:133-168.

Georghiou, G. P. (1972b) Studies on resistance to carbamate and organophophorous insecticides in *Anopheles albimanus*. Am. J. Trop. Med. Hyg. 21:797-806.

Georghiou, G. P., V. Ariaratnam, and S. G. Breeland (1972) Development of resistance to carbamates and organophosphorous compounds in *Anopheles albimanus* in nature. Bull. WHO 46:551-554.

Georghiou, G. P., S. G. Breeland, and V. Ariaratnam (1973) Seasonal escalation of organophosphorus and carbamate resistance in *Anopheles albimanus* by agricultural sprays. Environ. Entomol. 2:369-374.

Glass, E. H., R. J. Smith, I. J. Thomason, and H. D. Thurston (1972) Plant Protection Problems in Southeast Asia. UC/AID Pest Management Project Rept.

Hamon, J. (1970) Conference on *Anopheles* biology and malaria eradication. Misc. Publ. Entomol. Soc. 7(1):53.

Hannah, S. A. (1973) Progress, problems, and people. War on Hunger 7(2):13-18.

Lancaster, J. L., and N. P. Tugwell (1969) Mosquito control from applications made for control of rice water weevil. J. Econ. Entomol. 62:1511-1512.

MacDonald, W. W. (1972) Symposium on Vector Control and the Recrudescence of Vector Borne Disease. June 15, 1971. PAHO Sci. Publ. No. 238. Washington, D.C.: Pan American Health Organization. p. 44.

Mitchell, C. J., P. S. Chen, and T. O. Kuno (1972) Preliminary studies on the control of Japanese encephalitis vectors in Taiwan (China) by chemical means. Unpublished document. WHO/VBC/72.377. Geneva.

Nave-Rebollo, O. (1971) Consejo Centroamericano de Salud Publica. 9th Annual Public Health Congress. San Salvador.

Olszyna-Marzys, A. E., M. de Campos, M. Taghi Farvar, and M. Thomas (1973) Chlorinated pesticide residues in human milk in Guatemala (in Spanish with English summary). Bol. Ofic. Sanit. Panam. 74(2):93-107.

Paddock, W. C. (1970) How green is the green revolution? BioScience 20:897-902.

Pal, R. (1962) Contributions of insecticides to public health in India. World Rev. Pest Control I(2):10.

Pal, R. (1974) The present status of insecticide resistance in anopheline mosquitoes. J. Trop. Med. Hyg. 77:28-41.

Pan American Health Organization (1972) Status of Malaria Eradication in the Americas. XX Report. Washington, D.C. p. 122.

Reich, G. A., J. H. Davis, and J. E. Davies (1968) Pesticide poisoning in South Florida; An analysis of mortality and morbidity and a comparison of sources of incidence data. Arch. Environ. Health 17:768.

Sasser, J. N., H. T. Reynolds, W. F. Megitt, and T. T. Herbert (1972) Crop Protection in Senegal, Niger, Mali, Ghana, Nigeria, Kenya, Tanzania, and Ethiopia. UC/AID Pest Management Project Rept.

Scrimshaw, N. S., C. E. Taylor, and J. E. Gordon (1968) Interactions of Nutrition and Infection. World Health Organization Monograph Series No. 57.

Self, L. S., H. K. Shin, K. H. Kim, K. W. Lee, C. Y. Chow, and H. K. Hong (1971) Ecological studies on *Culex tritaeniorhynchus* in the Republic of Korea in 1971. Unpublished document. WHO/VBC/71.332.

Smith, R. F. (1972) Management of the environment and insect pest control. *In* "Ecology in Relation to Plant Pest Control." Rome: Food and Agriculture Organization.

Tapía, B. H. (1972) 16 anos de labor experimental del programa de mejoramiento de maíz en Nicaragua. Centro Experimental Agropecuario "La Calera." Ministerio de Agricultura y Ganadería, Managua, Nicaragua. Boletin Técnico No. 1.

Vandekar, M., R. Plestina, and K. Wilhelm (1971) Toxicity of carbamates for mammals. Bull. WHO 44:241.

Wolfenbarger, D. A., M. J. Lukefahr, and H. M. Graham (1971) A field population of bollworms resistant to methyl parathion. J. Econ. Entomol. 64(3):755-756.

World Health Organization (1970) 15th Report Expert Committee on Malaria. WHO Tech. Rep. Ser. No. 467, p. 20.

World Health Organization (1971) Insecticide resistance: The problem and its solution. WHO Chron. 25:214-218.

World Health Organization (1973) Safe Use of Pesticides. 20th Report, Expert Committee on Insecticides. WHO Tech. Rep. Ser. No. 513, p. 42.

Wright, J. W., R. F. Fritz, and J. Haworth (1972) Changing concepts of vector control in malaria eradication. Annu. Rev. Entomol. 17:75-102.

APPENDIX A:
RICE PRODUCTION IN ASIA AND ITS RELATIONSHIP
TO PUBLIC HEALTH

The study team made a detailed analysis of the transfer of one set of agricultural technologies, involving rice production, and its effect on the public health in Asia, primarily Southeast Asia. The acreage devoted to rice culture in Asia is immense, and many mosquito species have a known proclivity for using rice fields as habitats in the immature stages. Thus, it is reasonable to believe that anything that increases the extent of rice fields or changes the techniques used in their culture might have an effect on mosquito breeding and, hence, on human health.

There are several ways in which new rice production technology might affect vectors of disease. First, an increased use of insecticides for rice insect control might hasten the development of resistance in mosquitoes of public health importance, as has happened in Central America with cotton insecticides. On the other hand, the use of rice insecticides might also decrease the population of those mosquitoes, at least temporarily. Second, it seems inevitable that an increase in irrigated rice acreage, or an extension of the flooded period of existing acreage, will increase mosquito breeding sites.

More than 50 years ago, the relationship of rice culture to mosquitoes and malaria appeared to be in little doubt, either here or abroad. Hardenburg (1922) states that, in the United States, "an area within or near a rice field is the worst possible site for a home from an antimosquito point of view," and there is "no really effective means of protection in such a case." Abroad, he states that "little rice is raised in Ceylon because of the increased incidence of malaria which it gives rise to." Again, "attempts to control malaria about the rice fields of Italy have been given up," and "cultivation of rice is now forbidden within two kilometers of a village."

In view of these potential and actual interactions and the uncertainty concerning them, it seems worthwhile to consider the following questions:

- How is rice grown in Asia? To what extent has the Green Revolution changed former production patterns, especially by increasing the use of insecticides in rice culture, changing water manipulation techniques, or increasing irrigated acreage?
- What is the precise relationship of rice fields to arthropod-borne disease in Asia?
- What actual or apparent effects have new rice production technologies had on disease, and what is their potential impact in the future?

AGRONOMIC ASPECTS OF RICE PRODUCTION

In 1972, the production of rice in Asia* was estimated at 261 million metric tons, or 91 percent of the world production of 286 million metric tons. Table A-1 shows the area harvested, yield, and production of the different Asian countries.

Clearly, rice is grown in a great number of areas, with widely varying climatic and soil conditions. Rice production practices also vary considerably, not only among the rice-producing countries, but within each country as well.

To gain some conception of how rice is produced in at least a part of Southeast Asia, some of the agronomic practices used in the Philippines and Indonesia are summarized below.

Philippines

Rice was grown on approximately 3.3 million hectares in the Philippines in 1972, with an estimated production of 4.9 million metric tons. Most of the farms are small, about half of them smaller than 2 hectares. Rice yields vary greatly according to region, with Central and Northern Luzon averaging 50-100 percent higher yields than Visayas or Mindanao. Yields also depend on whether the crop is

---

*"Asia" includes the countries listed in Table A-1. Much of the discussion in this section refers primarily to Southeast Asia. However, examples are drawn from other parts of Asia as well.

Table A-1 Rice in Asia: Area Harvested, Yield, and Production in Specific Countries, 1972 (all figures are preliminary)

Area	Area Harvested (ha × 1,000)	Yield (quintals/ha)	Production (metric tons × 1,000)
Southeast Asia			
Burma	5,020	15.1	7,600
Cambodia	1,900	10.1	1,927
Indonesia	6,018	25.3	20,270
Laos	911	9.0	820
Malaysia	588	29.5	1,736
Philippines	3,324	14.6	4,854
Thailand	7,120	15.3	10,909
N. Vietnam	2,700	22.2	6,000
S. Vietnam	2,600	20.5	5,333
East Asia			
China, People's Rep. of	31,500	31.1	98,000
China, Rep. of (Taiwan)	742	42.3	3,137
Japan	2,643	56.3	14,871
Korea	1,200	44.5	5,342
Southern Asia			
Afghanistan	238	24.2	575
India	35,000	15.8	55,472
Nepal	1,200	20.0	2,400
Pakistan	11,013	16.6	18,292
Sri Lanka (Ceylon)	699	21.0	1,469

SOURCE: USDA, FAS, Washington, D.C. Foreign Agriculture Circular, August 1973.

grown as irrigated lowland, nonirrigated lowland, or upland rice. The varieties chosen and cultural practices used also affect yield.

Crop production begins with land preparation. The soil is usually prepared by using animals (caribou or water buffalo) to pull the various plows and harrows that destroy the existing rice stubble or straw and level the field for uniform irrigation and water distribution. Most of the acreage is plowed at the beginning of the rainy season, diked to impound and conserve the water, and puddled up to a fine consistency for transplanting, which is the most common method of planting lowland rice. The water softens the land and facilitates the various plowing and harrowing operations.

A seedbed is started 4-6 weeks before the beginning of the flood season. The seedlings are pulled 20-30 days after planting and then transplanted into the prepared field in hills of 2-4 seedlings each. Different methods of transplanting are practiced in the Philippines, but a substantial percentage of the lowland rice is planted in a random fashion. This method creates irregular distances between plants and makes the use of a rotary weeder almost impossible. The straight-row method of transplanting, with spacing adequate to permit weeding and other field operations, is practiced by some rice farmers.

A very few of the Philippine rice growers use direct seeding by drilling in dry soil, broadcasting on mud, broadcasting in water, or drilling in mud. The increased cost of seed and fertilizers, plus the success of transplanting, limits this practice.

Most of the rice is irrigated from the natural accumulation of water during the rainy season, June through November in most areas. Two methods of irrigating lowland rice are used: continuous submergence and intermittent irrigation and drainage. The conventional method, continuous flooding, maintains the water at a depth of 5-10 cm until the grain has developed fully. Intermittent irrigation consists of a series of cycles of draining and flooding--for example, 10 days flooded followed by 5 days drained. Application of fertilizers and weeding operations are usually carried out during the drained periods.

About one-fourth of the rice is planted in upland areas, where flooded conditions cannot be imposed. Most of the upland crop is planted on rolling to hilly soils by broadcasting the seed after land preparation. The weeding operation and the effective use of fertilizers is made more difficult because of the lack of irrigation water.

When fertilizers are available, nitrogen, phosphorus, and potassium are used to increase grain yields. When farmers cannot obtain commercial fertilizer at a reasonable cost, green manure crops and composts are usually used as sources of nitrogen.

Weed control currently requires a major labor input in rice production in the Philippines. For example, the labor distribution in Laguna province of Luzon island was estimated at 35 percent for weeding and only 1 percent for fertilization and insect control together. (Other labor requirements included soil preparation, 16 percent; transplanting, 11 percent; and harvesting, 37 percent.) It is hardly surprising that chemical herbicides have met with better acceptance than other pesticides. Herbicides are not yet widely available, however. Other methods of weed control include hand pulling, thorough land preparation, and control of the depth of irrigation water.

Destructive rice insects, as well as insect vectors of rice virus diseases, are controlled with chemicals where available. As with herbicides, however, insecticides are not available to a great number of small farmers because of distribution problems and costs.

Rice-field rats, which damage rice at all growth stages, are another serious problem in the Philippines. Control attempts include the use of rodenticides and mechanical traps and the reduction of food source and weed cover at the borders of the fields.

The irrigation water is drained 7-10 days before harvest. At this time, the grains are in the hard dough stage. Drainage hastens maturity and also improves harvesting conditions.

The crop is ordinarily cut by hand, bundled and allowed to dry in the field. It is then hauled to a threshing area where it is either threshed by hand or by small stationary threshing machines. This method of hand harvesting (cutting with a sickle; piling into stacks; collecting into large bundles; hauling to a threshing area; and manually threshing, winnowing, and sacking) requires about 250-350 man-hours of labor per hectare of rice. It is estimated that 950 man-hours are required all together to grow and harvest 1 hectare of rice, which yields an average of 1,500-2,100 kilograms.

Indonesia

Indonesia is an archipelago of more than 13,000 islands, of which about 1,000 are inhabited. The total land area of the country is approximately one-fourth that of the continental United States. The large islands of Java, Sumatra, Kalimantan, Celebes, and West Irian include more than 90 percent of the total land area.

In 1970, the population of Indonesia was about 120 million and growing at the rate of 2.4 percent per annum. More than two-thirds of the total population live on the island of Java, which contains only 7 percent of the land area. Java is one of the most densely populated areas of the world, with about 6 persons per hectare. Practically every hectare of usable land in Java is under cultivation, and some swamp areas are in the process of being reclaimed.

More than 70 percent of Indonesians are directly engaged in agriculture. Most of the cultivated land is divided into very small plots, with most farmers operating less than 1 hectare. Production from the native farms is used for domestic consumption. Despite recent improvements in agricultural technology, Indonesia must still import some rice.

In 1972, rice was grown on approximately 8.0 million hectares in Indonesia. Approximately 60 percent of the rice is grown in Java.

Rice is planted and harvested during all months of the year. The peak period is in the wet monsoon season (November to April) because of the need for water. About 85 percent of the rice is grown in association with irrigation, and 15 percent is grown as upland rice. Of the irrigated rice, 20 percent is under highly developed, gravity-flow irrigation systems developed by the Dutch engineers; another 20 percent is provided with water by simpler gravity-flow irrigation systems from local streams; and the remaining 60 percent is irrigated by the collection of natural rainfall within bunds. Expansion of rice production in Java is limited by lack of water during the dry season and by land with slopes too great to develop good irrigation systems.

Transplantation of seedlings is practiced in almost all of the irrigated rice fields in Java. Some direct seeding by hand broadcasting is used for the upland crop. Much of the rice in Java is grown on hillsides, and numerous terraces have been constructed. The individual fields are extremely small (much less than 1/4 hectare), and each farmer is dependent on the adjacent rice plot for water use and water control. The lowland rice fields are slightly larger but also tend to be restricted because of the hilly terrain.

Where it is feasible, water buffaloes are used for plowing and harrowing. In other areas, particularly on mountain slopes, the cultivation is done by men using hoes or picks. After the land is prepared, seedlings are transplanted into fields that are partially irrigated and set in clumps of 3-5 plants approximately 10 inches apart.

Women usually do the transplanting. Frequent weeding is required during the season.

Rice can be planted or harvested in any month of the year, subject only to the availability of water for irrigation. Farmers grow two rice crops a year on land where there is sufficient water. As a result, rice is harvested throughout the year, and insects and diseases are numerous and troublesome.

In 1963, the Indonesian government began a program designed to bring more advanced agricultural technology to Java. Farmers were offered a package of inputs on the basis of group credit against collective collateral security. These block credits were used to subsidize inputs of fertilizers, insecticides, and sprayers and also included some living costs for the farmers. Some of the programs also offered seeds of high-yielding rice varieties. In cooperation with the Indonesian government, many foreign countries provided additional inputs to designated blocks of farmers by supplying fertilizer, insecticides, spraying service, cost-of-living allowances, and extension service. These attempts to encourage advanced technological inputs have met with varying degrees of success. Spraying vast areas with ultra-low-volume insecticides or the use of small hand sprayers for application of chemicals for insect control has not been uniformly successful. Lack of adequate data for disease and insect control, fertilizer applications, and varietal performance have further complicated the success of these programs. The use of fertilizers and pesticides in Java was extremely variable in 1972-1973.

Rice is usually harvested in Java by cutting each individual head, tying these in bundles of 15 to 30 pounds, and then storing the bundles for later threshing by hand (or foot) when it is required for home consumption or for sale to the miller as an unthreshable bundle. The rice mills then thresh the rice and mill it. These bundles, called *paddy*, usually yield about 50 percent of milled rice (because of the stems) in contrast to 60-68 percent in other countries.

EFFECTS OF THE GREEN REVOLUTION

In the preceding section, the rice production practices in two countries in Southeast Asia were described in some detail. The difficulty in drawing generalizations about how rice is grown should be clear.

The several references to the introduction of new high-yielding varieties of rice and techniques for growing them should make it clear, as well, that the new technologies have met with uneven success. At the same time, to the extent that they have been accepted, they represent a major and rapid technological transfer from the industrialized to the less-developed countries. The transfer involves not only the seed-fertilizer technology itself, but also the *capacity* to produce new technological variation. It is the current and potential impact of this transfer, the longer-run and perhaps less obvious consequences of the green revolution, that concern us here. The following should help to clarify the nature of the technological transfer that has taken place.

What Is the Green Revolution?*

Although "green revolution" is a well-known term, it suffers from great impreciseness. Moreover, its emotional connotations have led to excesses in popular use--to overselling and overbuying. Nonetheless, the phrase is probably here to stay, even though the process it describes may be more evolutionary than revolutionary.

To remove some imprecision, a broad working definition is in order. The green revolution can be thought of as the adoption in less developed countries of high-yielding grain varieties--mainly rice and wheat--and the assorted package of inputs--mainly fertilizer, pesticides, and water control. The new varieties are characterized by short, stiff stalks; are fertilizer-responsive and photoperiod-insensitive; and are being improved steadily through continued breeding and experimentation. Hence, the green revolution is not a single event, but rather a package of evolving technology. Moreover, the required package may vary substantially among regions, depending upon water control and other environmental conditions.

What Is the Extent of the Green Revolution?

"High-yielding variety" is a term almost as ambiguous as "green revolution." As a consequence, it is difficult to

*Drawn heavily from Dalrymple and Jones 1973. (This report contains an excellent bibliography on recent materials on the green revolution.)

date the exact origins of the new technology or to delineate its precise acreage. Early-maturing varieties were known in China as early as 1000 A.D., and semidwarf wheats were noticed in Japan in 1873. The "modern" high-yielding varieties, however, were released in Mexico in the mid-1940s, and within 12 years, 90 percent of Mexico's wheat land was being planted to them. In other countries, the "new" high-yielding varieties date from approximately 1965/1966. In the case of rice, about 40 million acres were under the new technology by 1972/1973, the last year for which there are even crude data (see Table A-2). In Asia, use of the new varieties was still concentrated in a few countries, with India, Pakistan, Bangladesh, Indonesia, and the Philippines accounting for most of the improved rice acreage (Table A-3). In terms of area, the green revolution affected about 20 percent of total rice acreage in noncommunist Asia during 1972/1973 (Table A-4).

What Is the Nature of the Technology Package?

The technology package for rice typically has several specific elements, although it should be stressed again that packages vary by regions. For areas with good water control, short, stiff-strawed varieties have been developed that permit much heavier fertilization than did the older varieties, which grew tall and lodged if applications of nitrogen fertilizer exceeded about 40 pounds per acre. The new varieties accommodate applications of three times this amount of nitrogen, and the resultant paddy yields have ranged generally from 25 to 100 percent higher. The key point, however, is that yields from the new varieties *without* increased fertilizer are only marginally higher (and sometimes lower) than from traditional varieties. From a pest control point of view, the heavier foliage that results from the seed-fertilizer combination often provides a better environment for a number of important insects and diseases.

A second aspect of the new rice technology involves the length of the growing period. Generally, the new varieties require fewer days to mature; and therefore, an extra crop may sometimes be grown during the year. An essential aspect of this alteration in cropping patterns has been the daylight insensitivity that has been bred into the new varieties, i.e., their ability to mature during periods of varying amounts of daylight.

Where supplemental water is not available, a second crop is not usually grown, so the length of time standing

TABLE A-2 Estimated High-Yielding Wheat and Rice Area, Asia and North Africa, 1965/66 to 1972/73[a]

Unit of Area and Year	Area		
	Wheat	Rice[b]	Total
Hectares			
1965/66	9,300	49,400	58,700
1966/67	651,100	1,034,300	1,685,400
1967/68	4,123,400	2,605,000	6,728,400
1968/69	8,012,700	4,706,000	12,718,700
1969/70	8,845,100	7,848,800	16,693,900
1970/71	11,344,100	10,201,100	21,545,200
1971/72	14,083,600	13,443,400	27,537,000
1972/73 (Prelim)	16,815,500[c]	15,658,600[d]	32,474,100
Acres			
1965/66	22,900	122,100	145,000
1966/67	1,608,800	2,555,900	4,164,700
1967/68	10,188,800	6,437,200	16,626,000
1968/69	19,799,700	11,628,000	31,427,700
1969/70	21,856,400	19,394,300	41,250,700
1970/71	28,031,400	25,207,000	53,238,400
1971/72	34,800,500	33,217,900	68,018,500
1972/73 (Prelim)	41,551,100[c]	38,692,000[d]	80,243,100

[a] Excludes Communist nations.
[b] Excludes Japan and Taiwan.
[c] Includes Turkey at 1971/72 level.
[d] In addition, about another 429,600 hectares (1,061,400 acres) of rice were planted in Latin America (excluding Cuba).

SOURCE: Dalrymple 1974.

water is left on the field is actually reduced. Where irrigation water is available, however, an additional rice crop may mean standing water for more days during the year, and hence a better environment for certain pests, including those of public health importance (malaria, filariasis, schistosomiasis). Similarly, the greatly increased profitability of rice as a consequence of technological changes may induce a substitution of rice production for other "dry" crops, again involving increased aquatic conditions. Still another side effect occurs because of the additional time flexibility. As neighboring

TABLE A-3 Estimated Area Planted to High-Yielding Varieties of Rice in Asia

Country	Unit	1965/66	1966/67	1967/68	1968/69	1969/70	1970/71	1971/72	1972/73 (Prelim.)
Asia									
1. Bangladesh	Hectares	--	200	67,200	152,200	263,900	460,100	623,600	1,069,600
	Acres	--	500	166,000	376,000	652,000	1,137,000	1,541,000	2,643,000
2. Burma	Hectares	--	--	3,400	166,900	143,000	190,900	185,100	199,200
	Acres	--	--	8,500	412,400	353,300	471,800	457,300	492,200
3. India	Hectares	7,100	888,400	1,785,000	2,681,000	4,343,500	5,589,200	7,411,400	8,639,100
	Acres	17,650	2,195,200	4,410,700	6,624,800	10,732,700	13,811,000	18,313,600	21,347,200
4. Indonesia	Hectares	--	--	--	198,000	826,000	913,000	1,338,000	1,521,000
	Acres	--	--	--	489,000	2,041,000	2,256,000	3,306,000	3,758,000
5. Korea	Hectares	--	--	--	--	--	--	2,700	187,000
(South)	Acres	--	--	--	--	--	--	6,700	462,000
6. Laos	Hectares	--	360	1,200	2,000	2,000	53,600	30,000	50,000
	Acres	--	900	3,000	5,000	5,000	132,500	74,100	123,600
7. Malaysia	Hectares	42,300	62,700	90,700	96,100	132,400	164,600	196,900	217,300
	Acres	104,450	155,000	224,200	237,500	327,100	406,600	486,500	537,000
8. Nepal	Hectares	--	--	--	42,500	49,800	67,800	81,600	177,300
	Acres	--	--	--	105,100	123,000	167,600	201,700	438,000
9. Pakistan	Hectares	--	80	4,000	308,000	501,400	550,400	728,500	643,500
	Acres	--	200	10,000	761,000	1,239,000	1,360,000	1,800,000	1,590,000
10. Philippines	Hectares	--	82,600	653,000	1,012,800[a]	1,354,000	1,565,000	1,827,000	1,752,000[a]
	Acres	--	204,100	1,613,600	2,500,000[a]	3,345,700	3,867,100	4,514,500	4,329,200[a]
11. Sri Lanka	Hectares	--	--	--	7,000	26,300	29,500	29,600	17,600
	Acres	--	--	--	17,200	65,100	73,000	73,100	43,500
12. Thailand[a]	Hectares	--	--	--	--	5,000	115,000	315,000	350,000
	Acres	--	--	--	--	12,400	284,000	778,000	865,000
13. Vietnam	Hectares	--	--	500	40,500	201,500	502,000	674,000	835,000
(South)	Acres	--	--	1,200	100,000	498,000	1,240,400	1,665,400	2,063,300
Total Asia	Hectares	49,400	1,034,300	2,605,000	4,706,000	7,848,800	10,201,100	13,443,400	15,658,600
(rounded)	Acres	122,100	2,555,900	6,437,200	11,628,000	19,394,300	25,207,000	33,217,900	38,692,000

[a]Unofficial estimate.

SOURCE: Dalrymple 1974.

TABLE A-4 Proportion of Total Rice Area Planted to High-Yielding Varieties

Country	Crop Year 1965/66	1966/67	1967/68	1968/69	1969/70	1970/71	1971/72	1972/73
	Percent							
Asia								
Bangladesh	--	negl.	0.7	1.6	2.6	4.6	6.7	11.1
Burma	--	--	negl.	3.3	2.9	3.6	3.6	4.2
India	negl.	2.5	4.9	7.3	11.5	14.9	19.9	24.7
Indonesia	--	--	--	2.4	10.3	11.2	15.9	18.0
Korea (South)	--	--	--	--	--	--	0.2	15.6
Laos	--	negl.	0.2	0.3	0.2	6.0	3.3	5.5
Malaysia	10.0	14.7	20.6	20.1	26.4	30.9	35.7	38.0
Nepal	--	--	--	3.7	4.4	5.8	6.3	14.8
Pakistan	--	negl.	0.3	19.8	29.9	36.6	50.0	43.4
Philippines	--	2.7	19.8	30.4[a]	43.5	50.3	56.3	56.3[a]
Sri Lanka	--	--	--	1.0	3.9	4.4	4.2	2.5
Thailand[a]	--	--	--	--	0.1	1.5	4.0	4.9
Vietnam (South)	--	--	negl.	1.6	8.3	19.9	25.9	32.1

[a]Based on unofficial estimates of HYV area.

SOURCE: Dalrymple 1974.

farmers choose alternative planting and harvesting dates, certain pest control methods (e.g., aerial spraying) may become less effective, and certain pest problems may become more acute (e.g., rats concentrating on and attacking early or late fields).

A third significant element of the technology involves water control. The best success with the new varieties has been in areas with technical irrigation. Relatively little success has occurred to date with new rice varieties in dry-land or deep-water areas, although current research efforts appear to hold substantial promise for these regions. Under rainfed monsoon conditions (where deep flooding is not a problem) the results have been mixed. There is a strong interaction between sunlight, fertilization, and water (see Figure A-1), and although many of the new varieties are indeed planted under rainfed conditions, they do not so clearly dominate the older varieties as they do in the well-irrigated areas such as the Indus basin.

The fourth element of the technology involves pesticides. It is much more difficult to generalize about the need for, or the effects of, pesticides than for fertilizer. Many of the traditional varieties were selected for their disease resistance, and the expectation has been that pesticides were "critical" to the new technology. Moreover, most extension recommendations include specific pesticide guidelines based on experiment-station evidence that shows a large physical increase in output as a result of their application (see Table A-5 for a typical report on Indonesia's testing). At the same time, the accumulated field evidence indicates that most farmers are not using pesticides on improved varieties, and in fact, less pesticide may be used now than was used a few years ago. Whether this relatively short-run "neglect" will precipitate longer-run problems remains to be seen. Whether, also, there will be increased ecological difficulties from the use of one or two varieties is a hotly debated topic. With a large number of traditional varieties, each with varying tolerances to different insects and diseases, there was thought to be some "natural control" in effect. With only one or two varieties, the fear is that production in whole regions might be quickly destroyed by an attack. Perhaps the widespread outbreak of rice tungro virus in the Philippines gives credence to such a concern.

On the other hand, a major focus of the breeding program at the International Rice Research Institute (IRRI) has been on disease resistance. The early IRRI varieties,

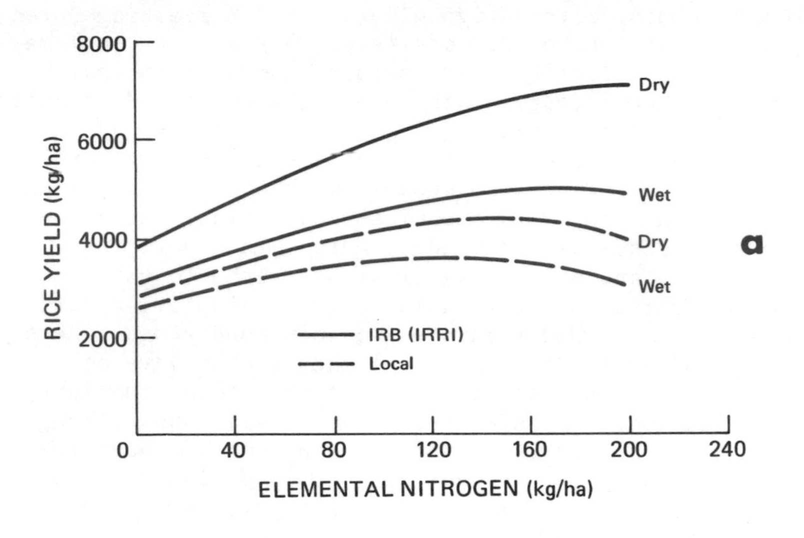

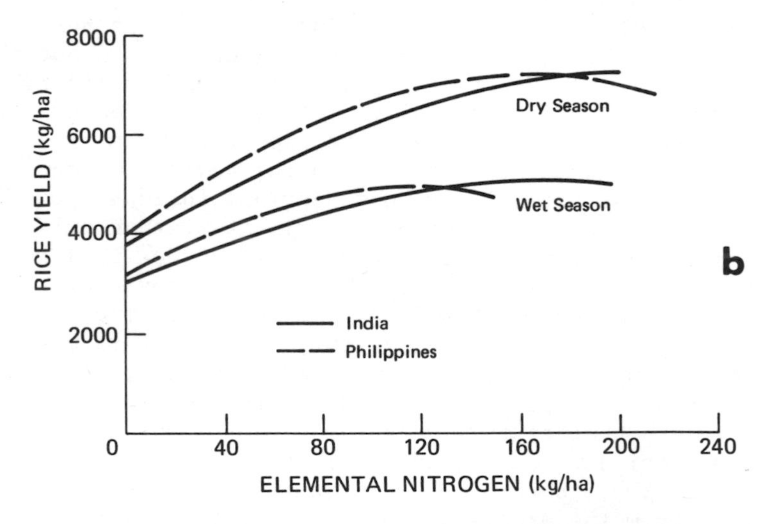

FIGURE A-1. Interaction of sunlight, water, and yields per hectare. a. Variability in yield response of rice to nitrogen by variety and season. India, 1968. Based on experiments of All India Coordinated Rice Improvement Project. b. Variability in yield response to IR8 variety to nitrogen by location and season. India, 1968 (19 locations) and Maligaya Rice Experiment Station in Central Luzon, Philippines (average of 1966, 1967, and 1968). Source: International Rice Research Institute 1971b. Rice Policy Conference: Current Papers from the Department of Agricultural Economics. Los Baños. May 1971.

TABLE A-5 Results of Pesticide Experiments on Rice Yields in Indonesia

Product	Total Dosage in kg Active Ingredient/ha	65 days after planting % Silver-shoots	% Dead Hearts	At harvest % White Heads	Yield (kg/ha)
Endrin 19,2ec	1.6	1.2	9.5	3.9	5,451
Endrin 2G	5	0.5	5.1	3.4	5,362
Sandoz 6626 5G	5	0.5	15.7	4.6	5,109
Solvigam 5G	5	0.2	22.9	6.2	5,106
Gusathion 40ec	2.4	0.7	12.1	3.3	4,957
Sevidol 4-4G	5	0.4	13.3	6.7	4,851
Birlane 5G	5	0.1	19.7	1.4	4,783
Thiodan 5G	5	0.9	16.9	4.9	4,780
BHC-6G	5	0.2	20.8	3.2	4,766
Surecide 25ec	2.4	0.8	18.4	3.5	4,597
Thiodan 35ec	2.4	2.9	12.9	5.9	4,178
Salithion 10G	5	0.3	19.4	7.0	4,156
Diazinon 60ec	2.4	0.7	26.1	8.0	4,154
Dimecron 100	3.2	1.3	10.7	8.0	4,147
Diazinon 10G	5	0.3	20.0	7.5	4,111
Birlane 24ec	2.4	0.6	34.0	6.1	3,978
Sumithion 50ec	3.2	0.5	16.5	8.0	3,847
Control	--	0.5	42.4	9.6	2,805
			MSD - 5%		1,493
			1%		1,725

SOURCE: International Rice Research Institute, 1971a.

such as IR5 and IR8, were susceptible to a variety of disease, insect, and soil problems. As shown in Figure A-2, however, the later varieties, such as IR26, have substantially more resistance. To the extent that insect and disease problems can be solved at the breeding stage, the need for pesticides can be substantially reduced.

What Are the Factors Limiting the Spread of the Green Revolution?

The new rice varieties seem to be more limited than is realized by many who talk about the green revolution. The constraints arise from both physical and economic factors.

Of the physical impediments, water control seems to be most severe. Most of the rice in South and Southeast Asia is grown in regions of high seasonal rainfall; only about 20 percent of the rice is irrigated, and not even all of this area is under complete water control. In Thailand, perhaps only 10 percent of the rice area has the degree of water control necessary for optimal results, and in India, none of the high-yielding varieties introduced to date are fully adapted to the monsoon conditions. This does not mean, however, that new varieties are not sown during the rainy season. Seventy percent of the new rice varieties grown in India were planted under these conditions in 1968, and in the Philippines, more area has been planted in the wet than the dry season. The conclusion is that, without good water control, yields of new varieties are more variable than those of traditional strains (Dalrymple and Jones 1973).

Another important constraint in some areas is the farm delivery system for key inputs such as fertilizer and pesticides. It is rare to find a country in Asia where physical availability of fertilizer has not been a severe problem sometime during the past several years. Moreover, with the exception of Taiwan, none of the countries of South and Southeast Asia has a widespread, fast-reacting, research-backed program on pesticides. As a consequence, local varieties sometimes have a superior resistance or may require much less in the way of out-of-pocket costs on disease control.

"Economics" is a third factor limiting the expansion of new varieties. The physical factors indicated above may mean that the new varieties are "uneconomic," but various government policies have their impact, too. For example, there is an enormous difference in rice prices

		Diseases					Insects			Soil problems			
	Lodging	Blast	Bacterial blight	Bacterial leaf streak	Grassy stunt	Tungro	Green leaf-hopper	Brown plant-hopper	Stem borer	Alkali injury	Salt injury	Iron toxicity	Reduction products
IR8	R	MR	S	S	S	S	R	S	MS	S	MR	S	MR
IR5	MR	S	S	MS	S	S	R	S	S	S	MR	S	MS
IR20	MR	MR	R	MR	S	R	R	S	MS	S	MR	R	MR
IR22	R	S	R	MS	S	S	S	S	S	S	S	MR	MR
IR24	R	S	S	MR	S	MR	R	S	S	MR	MR	MR	MS
IR26	MR	MR	R	MR	MR	R	R	R	MR	MR	MR	R	MR

FIGURE A-2. Susceptibility of IRRI varieties to disease, insect, and soil problems. S, susceptible; MR, moderately resistant; R, resistant. Source: International Rice Research Institute, 1974.

(and in rice-to-fertilizer price ratios) among Asian countries (see Table A-6). It is not a complete coincidence that those countries with high yields have favorable economic climates (Table A-7). (On the other hand, there are important differences in yield potentials among temperate and tropical countries, and even with identical economic policies, the tropical areas would do less well in terms of yields.)

Who Has Participated in the Green Revolution, and Has It Been a Stabilizing or Destabilizing Influence?

While certain specifics of the foregoing points are arguable, there is fairly widespread agreement on them. This is not the case, however, with the equity/employment/stability issue. Because of its complexity, this point has become almost a philosophical matter on which there have emerged several distinct points of view. On the one side, the UN Global Two Study sees the Green Revolution in terms of increasing disparity among classes, regions, and nations, with these side effects offsetting much, perhaps all, of the production gains (e.g., Griffin 1972). By contrast, a group of country case studies conducted under IRRI auspices has emphasized the positive accomplishments of increased food production, of general participation by small farmers, and of increased employment. More equivocal positions are developed in studies by Wharton (1969) and Dalrymple and Jones (1973). Because of these differences, any (or perhaps no) opinion can be supported by the literature. The following is one assessment:

1. The green revolution was a significant factor in raising rice yields and production, without which most Asian nations would find themselves in even worse shape from a broad development perspective.
2. The green revolution has aggravated, in some instances severely, disparities among regions within countries. In particular, the irrigated regions, which were often the most prosperous anyway, have been the clear beneficiaries of the new technology. In this sense, the green revolution can be viewed as being a destabilizing influence.
3. The employment effects of the new technology must be defined carefully. Under the new system, more labor seems to be used per hectare (perhaps 30 percent more), but less labor per hundredweight of output (perhaps 15

TABLE A-6 Comparative Price Data for Fertilizer and Rice[a]

Country	Paddy Price to Producers (U.S. %/kg)	Price of Fertilizer Nutrients to Producers (U.S. ¢/kg)	Ratio of Paddy Price to Fertilizer Price	Paddy Yield in 1970 (MT/ha)
Japan	30.7	21.5[b]	1.428	5.64
South Korea	18.4 (I)	19.1	0.963	4.55
Taiwan	11.7[b]	26.2[b]	0.447	4.16
Malaysia	8.8 (I)	20.3[c]	0.433	2.72
Sri Lanka	11.3	15.8	0.715	2.64
Indonesia	4.5 (I)	15.2 (I)	0.296	2.14
Thailand	4.5[b,g]	14.3-50[h]	0.315- 0.090	1.97
Philippines	7.0 (I)[f]	17.3[f]	0.405	1.72
Burma	3.1[d]	25.1[e]	0.124	1.70

[a]Prices are from FAO *Production Yearbook 1971* unless otherwise indicated. A (I) indicates the source is the relevant paper in IRRI, *Viewpoints on Rice Policy in Asia*, 1971. In all cases attempts have been made to approximate prices actually received by producers.
[b]Average 1969/70 and 1970/71.
[c]Calculated from *Padi Farming in West Malaysia*, Malaysian Department of Agriculture, March 1972.
[d]Figure for 1965. Trade reports indicate the paddy to fertilizer ratio improved by 1970, so a figure of 0.20 was used in the analysis.
[e]Figure for 1969/70 only.
[f]The producer price of paddy was P0.45/kg. and the nutrient cost of fertilizer was P1.111/kg. The "fluctuating free exchange rate" of P6.43/$ was used for the conversion.
[g]Calculated from the price of #2 paddy in Bangkok multiplied by 0.9 to get a producer price.
[h]The fertilizer price is a 1967 price for ammonium phosphate. The low price assumes that both the N and P_2O_5 content contribute to increasing output; the higher price assumes only the N content is effective. All other fertilizer prices are for the N content of urea.

SOURCE: Timmer and Falcon 1975.

TABLE A-7 Asian Paddy Yields (kg/ha)

Country	1963	1964	1965	1966	1967	1968	1969	1970	1971	1972
India	1,550	1,610	1,300	1,300	1,550	1,610	1,610	1,700	1,710	1,648[a]
Pakistan								2,196[a]	2,300[a]	2,288[a]
Pakistan (undivided)	1,724	1,680	1,650	1,570	1,680	1,780	1,810			
Bangladesh								1,686[a]	1,522[a]	1,500[a]
Thailand	1,580	1,600	1,540	1,840	1,840	1,910	1,930	1,970	1,970	1,815[a]
Malaysia	2,570	2,320	2,620	2,560	2,620	2,650	2,710	2,720	2,950	2,900[a]
Indonesia	1,770	1,810	1,850	1,820	1,760	1,850	1,940	2,140	2,200	2,149[a]
Philippines	1,240	1,250	1,310	1,320	1,440	1,330	1,680	1,720	1,720	1,582[a]
Korea	4,400	4,470	3,850	4,300	4,060	3,860	4,660	4,550	4,580	4,618[a]
Taiwan	3,690	3,850	3,980	3,950	4,020	4,180	3,870	4,160	3,420 3,950[b]	--[a] 4,230[b]
Japan	5,090	5,010	4,950	5,090	5,750	5,720	5,550	5,640	5,250	6,059[a]

SOURCE: *FAO Production Yearbook*, 1971, and earlier issues.

[a]Calculated from production and area in *FAO Monthly Bulletin of Agricultural Economics and Statistics*, February 1973.

[b]U.S. Department of Agriculture *World Agriculture Production and Trade, Statistical Report*, June 1973. FAO no longer recognizes Taiwan.

percent less). It is in the latter sense that the new varieties are laborsaving.

4. There do appear to have been severe labor displacements in the case of rice arising from tractorization and/or consolidation of sharecropped units into one large farm. Hence, the experience with rice is rather different from that with wheat, where certain displacement problems have been locally severe.

5. Some of the indirect consequences have been large and unexpected. In Indonesia, for example, certain technical features of the new varieties have induced important changes in harvesting arrangements and techniques. The new surpluses have also given a boost to the local mechanical rice-milling industry, with a consequent decline in "hand pounding." The shift to more polished forms of rice is having, in turn, important negative nutritional consequences.

6. Small farmers have lagged in the utilization of the new technology. However, when they have not been excluded from the new inputs (e.g., through credit or fertilizer rationing), small farmers after 3 or 4 years seem to use the new technology in about the same proportion as their larger neighbors.

What Are the Implications of the Use of Insecticides in Asian Rice Fields?

To date, only a very low percentage of either large or small farmers appear to be using pesticides in the less developed countries of Asia. Some cases of excessive pesticide use can be documented. An example is the now abandoned Bimas Gotong Rojung scheme in Indonesia, where several large foreign firms contracted to supply key inputs, including aerial spraying, to small farmers. In general, however, the evidence seems to indicate that the use of pesticides on rice may even be declining rather than increasing.

Experimental data appear to show considerable yield increases as a result of insecticide application. In farmers' fields, however, it is virtually impossible to document the effects of spraying on output, as yields obviously depend on such factors as infestation levels and the number of farmers who do spray. In short, few farmers are using pesticides--perhaps because benefits are so hard to document--and as a consequence, it is very difficult to demonstrate public health effects either of a positive or negative nature.

In view of the fact that insecticides are not widely used in Asia, it is not surprising that there have been few documented cases of the development of resistance in disease vectors due to agricultural spraying. Chemical control programs against rice insects in some parts of Asia have been shown to have reduced vector populations, at least temporarily. This occurred in Korea with vectors of Japanese encephalitis.

DISEASE VECTORS ASSOCIATED WITH RICE FIELDS

The optimal use of high-yielding varieties of rice in parts of Asia depends on improved water control. In many regions this will necessitate expansion of systems for irrigation and drainage. These developments in turn may extend the habitat of various water-breeding vectors of disease. A detailed look at these organisms and their habitats is prerequisite to an evaluation of their potential for extending their ranges.

The three most important arthropod-borne diseases in Asia that may be associated with rice fields are malaria, filariasis, and Japanese encephalitis. A fourth disease associated with rice, schistosomiasis, is carried by snails. The precise relationship of these four diseases to their rice-field habitats is examined below.

Malaria and Filariasis

Malaria, caused by protozoan parasites in the red blood cells, is transmitted by the bite of mosquitoes of the genus *Anopheles*.*

---

*The life cycle of the malaria parasites (genus *Plasmodium*) involves both a sexual generation, which takes place in the mosquito host, and an asexual generation, which occurs in the human host. The parasite's immature macrogametocytes and microgametocytes can multiply rapidly in human blood but can mature and unite only when they are transferred with a blood meal from an infected person to the stomach of a female *Anopheles* mosquito. The union of gametocytes in the mosquito results in the formation of oocysts, which release minute sporozoites. These invade the mosquito's salivary glands, and pass into a new human host when the mosquito bites another person.

Unlike the situation in many parts of the world, the most intensive malaria transmission in Asia occurs in foothills, where *Anopheles minimus* and its close relatives, and in certain areas *A. leucosphyrus* and *A. balabacensis*, are the principal vectors. Along the coast where there are brackish marshes and salt-water fish ponds, *A. sundaicus* is of importance. Throughout most of India, the bulk of the malaria endemicity was maintained by the pool-breeding *A. culicifacies*. In general, the species associated with rice cultivation have not been responsible for epidemics or high endemicity of malaria. However, there are areas where certain of the rice field *Anopheles* have maintained a low but extensive and persistent endemicity. The most important of these species are listed in Table A-8.

Filariasis is caused by nematode parasites, primarily *Wuchereria bancrofti* and *Brugia malayi*, that live in the blood and lymph of man. Vectors include 20 or more species of *Aedes*, *Anopheles*, *Culex*, and *Mansonia* mosquitoes.* Filariasis in Southeast Asia is caused by periodic *W. bancrofti*† and is primarily a disease of urban areas, where *Culex quinquefasciatus* is the vector. However, in rural areas the *Anopheles* malaria vectors often also transmit *W. bancrofti*.

---

*The human acts as host to both adult and larval filarial parasites, but a part of the worms' larval development must take place within the vector mosquito. In the human host, the young larvae (called microfilariae) swarm into the superficial blood vessels at certain times (periodically or not), at which time mosquitoes may draw in some of the parasites with a blood meal. The worms develop into a different, infective larval form in the muscles of the mosquito. They may then bore out of the mosquito's mouth parts and into the skin of man as the mosquito is feeding. In man, the worms migrate, generally to the lymphatics, where they grow to a length of about 3 inches, mate, and produce young.

†There are two major forms of filaria, the periodic and the subperiodic. In the former, the microfilariae circulate periodically, passing through the peripheral blood vessels at night and retreating to the lungs and deeper blood vessels during the day. In the subperiodic form, the microfilariae are present in the peripheral blood by day and by night, but show some indication of a peak in the evening hours.

TABLE A-8 Rice-Field *Anopheles* As Malaria Vectors in Asia

Species	Geographic Area	Importance of Rice Habitat[a]	Importance As Vector[a]
A. aconitus	Java,	+++	+++
	Thailand	+++	+
A. annularis	Assam	+	+
A. barbirostris[b]	Celebes	+	+
A. campestris	West Malaysia	+	++
A. culicifacies	India	++	+++
A. fluviatilis	India	+	+++
A. minimus	Widespread	±	+++
A. nigerrimus	Celebes	++	+
A. philippinensis	India, Burma	+++	+
A. sinensis[b]	China, North Vietnam	+++	+++

[a]Subjective assessments, based on the literature, for the area in question.
[b]Taxonomic status requires clarification.

Filariasis caused by *Brugia malayi* exists in two forms. The subperiodic form affects both monkeys and man in the swamp forests of Malaysia and Palawan; it is transmitted by species of *Mansonia*. The periodic form occurs in areas of open rice fields in northwest Malaya. This form is also transmitted by *Mansonia*, and at times, by species of *Anopheles* as well. Table A-9 lists the areas where rice field mosquitoes act as vectors of filariasis.

Mansonia generally do not breed in rice fields. The role of the rice-field *Anopheles* in the transmission of both malaria and filariasis is discussed in detail below.

TABLE A-9 Rice-Field Anopheles Species As Vectors of Human Filarial Parasites in Asia

Mosquito Species	Parasite Species	Geographical Area	Importance of Rice Habitat[a]	Importance as Vector[a]
A. campestris	*Brugia malaye*	West Malaysia	+	++
A. flavirostris	*Wucheria bancrofti*	Philippines	±	+
A. annularis	*Wucheria bancrofti*	Philippines	++	+

[a]Subjective ratings, from the literature.

The *Anopheles hyrcanus* Group*

The distribution of *A. sinensis* (and possible close relatives) extends from northern China into the Malay peninsula and Sumatra. Chow (1949) states that in China it breeds most commonly in rice fields and that it is the chief malaria vector in the central plain of China, although the natural infection rates were only 0.1 to 0.3 percent. Hu (1935) showed that the highest seasonal density of *A. sinensis* was correlated with the highest prevalence of malaria in the Shanghai area. Hu and Yu (1937) refer to the abundance of the larvae in rice fields and irrigation ditches. Here *A. sinensis* adults fed by preference on large domestic animals; however, from 5 to 8 percent were positive for human blood, and this was enough to ensure the continued transmission of malaria.

In Korea, *A. sinensis* has been suspected of being the malaria vector on epidemiological grounds, and in 1962 one of 5,000 dissected was found with sporozoites. This again is an indication of a very low vector efficiency, but the high density of the species in rice fields permits a certain amount of transmission to take place. Chen et al. (1967) showed that malaria occurred in those localities where rice cultivation depended upon rainfall and the water had to be kept in the fields as long as possible. In areas with adequate irrigation systems where the rice fields were not continually flooded, there was no apparent malaria.

There are records of natural infections of *A. sinensis* from Java, Sumatra, Malaysia, and Southeast Asia (see Covell 1944).†

---

*This group of mosquitoes has undergone an intensive taxonomic study by Reid (1953) and more recently by Harrison (1972). Even so, the exact interspecific relationships do not seem to be entirely clarified. There are many references to malaria transmission by *Anopheles sinensis*, and although some of the forms now recognized as distinct species may have been included in the early studies on *A. sinensis*, probably the bulk of the work actually was done on the species now called *A. hyrcanus*.

†As Reid (1953) points out, it is impossible to determine exactly which of the presently recognized species were involved, but it seems probable these were *A. sinensis* and *A. nigerrimus*. The latter may have been a vector in some areas, but in general it appears to be of less

A. sinensis has been found in China naturally infected with the filarial parasite *Wuchereria bancrofti* (Chow 1950). In Malaysia, *A. sinensis* has been found naturally infected with periodic *Brugia malayi*, but at a very low rate of only 1 percent. Here *A. barbirostris* (= *campestris*) also was naturally infected (3 percent). The important vectors are species of *Mansonia*, with infection rates of 80 percent or more, so the *Anopheles* probably are of little epidemiological importance.

The *Anopheles barbirostris* Group*

A. campestris occurs in the broad alluvial plains along the coast and river deltas of Malaysia. There are records from India, Bangladesh, Burma, and Thailand; Reid (1962) suspects it is present in other parts of Southeast Asia as well. Reid describes the larval habitats as being deep water with some vegetation, in ditches, earth wells, and swamps; also the corners and edges of rice fields where the water is deep and shaded by overhanging trees. These places usually are near houses. Reid asserts the evidence is quite conclusive that *A. campestris* is an important vector of malaria on the western coastal plain of Malaysia. In almost 15,000 specimens dissected in the region, the average sporozoite rate was 0.33 percent.

In Malaysia, Wharton et al. (1964) found larvae resembling the filaria *B. malayi* in *A. donaldi*; the infective-larvae rate was 1.4 percent. Wilson (1961) points out that periodic *B. malayi* in Malaysia occurs most commonly in coastal rice fields interspersed with patches of open swamp forests. Here natural infections have been found in *A. barbirostris* as well as in *Mansonia indiana* and *M. uniformis*. As noted for malaria, the records for *A.*

---

importance than *A. sinensis*. *A. nigerrimus* also is found in rice fields.

*In his 1962 revision of this group, Reid has established several species that formerly were lumped under the name of *A. barbirostris*. The species now recognized as *A. barbirostris* is thought by Reid to occur in India, Ceylon, Southeast Asia, and Indonesia from Sumatra to Celebes and the Moluccas. The mosquito uses a variety of larval habitats, from deep-water ponds to rice fields. There are earlier records associating *A. barbirostris* with malaria, but it appears that these actually should refer to *A. campestris* and *A. donaldi*.

barbirostris probably actually refer to the sibling species *A. campestris*. Moorehouse (1962) reported that this rice-field breeder disappeared from coastal West Malaysia following several rounds of DDT spraying for malaria control.

Members of Other Species Groups

A. philippinensis is a widespread species with a range extending from Hainan and the Philippines through Indochina and about the eastern third of India, and including most of Indonesia. It breeds in quiet waters such as reservoirs, pooled streams, ditches, and rice fields. In the Bengal delta, where there are extensive swamps and "dead" rivers in which rice and other crops are grown, *A. philippinensis* finds suitable breeding places. The species was considered in former years to be an important vector in Bengal, but it seems to have lost whatever importance it may have had. Viswanathan et al. (1941) found naturally infected specimens in Assam in 1931 and 1933, but not thereafter. Neogy and Sen (1962) say, ". . . it is common knowledge that the chief vector in the area (West Bengal) i.e., *A. philippinensis* is gradually vanishing from alluvial deltan Bengal." Furthermore, it is not uncommon in some parts of the Philippines, where it has not been associated with malaria.

A. annularis also breeds in the extensive swamps and "dead" rivers in which rice is grown in the Bengal delta. In past years it was considered to be a malaria vector. In Assam, Viswanathan et al. (1941) found seven specimens with oocysts and four with sporozoites in 20,043 dissected. In certain localities it was the only species found infected. Rao (1949) says that in the plains of Orissa it was the main, if not the only, vector of malaria.

A. aconitus is thought to be an important malaria vector in Indonesia, perhaps the principal vector in inland areas of Java (Chow et al. 1960). The affected areas are those with rice fields. Here the species was largely exophagous, and although it fed mostly on animals, 17 percent of the femals taken in outdoor resting places contained human blood.

The most recent data from Indonesia continue to indicate an important vector role for *A. aconitus* and a growing problem of resistance by this species to DDT, dieldrin, and gamma-HCH (T. O'Connor, WHO Malaria Team, Personal communication, 1972).

Foci of malaria exist in the central plains of Thailand, where *A. minimus* and *A. balabacensis* are absent. Gould et al. (1967) dissected approximately 3,000 adult *Anopheles* that had been caught off human bait inside and outside houses. Of these, 473 were *A. aconitus*, and two of them were infected, one with oocysts and the other with both oocysts and sporozoites. The latter had bitten one of the collectors, who subsequently came down with a *P. vivax* infection.

Because *A. aconitus* is very widely distributed and abundant in Southeast Asia and is not an important vector other than in Java, one is led to suspect that a species complex is involved. From what has been seen in Thailand and elsewhere (Scanlon et al. 1968), the whole *A. minimus*, *aconitus*, *flavirostris*, *varuna* complex is a difficult tangle. However, the *A. aconitus* specimens examined in Java appear to be inseparable from the Thai specimens, at least in the female stage.

The finding by Gould et al. (1967) that *A. aconitus* was involved in a limited *P. vivax* epidemic in the rice plain north of Bangkok is difficult to interpret. If *A. aconitus* were a vector there with any degree of regularity, one would expect to see large outbreaks, since *A. aconitus* is often extremely abundant in the area. Very small pockets of malaria have been reported from this essentially malaria-free area from time to time, always quite limited. On at least one previous occasion the vector was said to be *A. philippinensis*.

Arthropod-Borne Viral Diseases

A number of these viral diseases have been investigated in Southeast Asia in recent years, and it is probable that others remain to be discovered. Among the most important from the public health viewpoint have been the several dengue strains and the hemorrhagic fever syndrome that frequently accompanies it. Because container-breeding *Aedes* species are the vectors of the dengue viruses, there is obviously no direct rice field association; it is an urban rather than a rural problem. The only important mosquito-borne virus in Asia associated with rice culture to some degree is Japanese encephalitis.

Japanese Encephalitis

Japanese encephalitis is an inflammation of the brain caused by a virus, transmitted primarily by mosquitoes of the genus *Culex*.* The disease has a distribution extending along the Pacific coast of Asia from Manchuria to Malaya and in the Pacific islands from Honshu to Java. Many birds can become infected with titers high enough to infest mosquitoes that feed on them, but mammals, especially horses and pigs, are also important as hosts.

The most important vector of Japanese encephalitis appears to be *Culex tritaeniorhynchus*. Experiments by Gresser et al. (1958) and many others have shown that this mosquito becomes infected after consuming very small quantities of virus. There have been numerous isolations of the virus from this species in nature. An example to illustrate the density that can be achieved by this species is that of Wada et al. (1967). In 1965, these authors collected 149,000 adult mosquitoes from 19 villages near Nagasaki City, of which 72.9 percent were *C. tritaeniorhynchus*. This mosquito breeds in many types of quiet waters and takes advantage of rice fields to build up high population densities.

A related species, which also includes rice fields among its larval habitats, is *Culex vishnui*.† It is an efficient vector of Japanese encephalitis virus and has been found naturally infected in India. Two other related species are *C. gelidus* and *C. annulus*; both have been found naturally infected, the former in malaya and Borneo (Gould et al. 1962, MacDonald et al. 1967) and the latter in Taiwan (Detels et al. 1970). These species are found in rice fields as well as in other waters. Thus, there is a group of banded-proboscis *Cules* species whose population densities are affected by the availability of rice fields, and because of these high densities they are sometimes responsible for outbreaks of encephalitis as well as for maintaining varying degrees of endemicity.

Culex fuscocephala, not a member of the banded-proboscis group, has also been implicated as a vector of Japanese encephalitis in the Chiengmai Valley of northern Thailand,

---

*The virus is ingested with a blood meal and multiplies within the mosquito. In about 10 days, the concentration of virus is high enough to allow transmission through the mosquito's salivary glands. If an infectious dose is transmitted to man, the disease develops. Other mammals and birds often show no apparent illness but may circulate the virus and reinfect large numbers of mosquitoes.

†This includes *C. pseudovishnui*, and possibly other species, as taxonomic work on this group is not complete as yet.

along with *C. tritaeniorhynchus* and *C. gelidus* (Muangman et al. 1973). *C. fuscocephia* is, if anything, more strongly correlated with rice culture than the banded-proboscis species, and it has also been implicated as a Japanese virus vector on Taiwan.

Because of the post-World War II experience of Korea with Japanese encephalitis, Korea and the World Health Organization have established several long-term studies of Japanese encephalitis and mosquitoes in that country. In parts of Korea, lowered *Culex* production in rural rice-growing areas was attributed at least in part to increased use of pesticides and more frequent flooding and draining.

Schistosomiasis

Of the three species of *Schistosoma* (blood flukes) that cause disease in man, only *S. japonicum* occurs in Asia. It is found in Japan, Formosa, China, the Philippines, Celebes, and Thailand. The adult female worms in the human host produce eggs that are passed on with the feces. The eggs are ingested by snails of the genus *Oncomelania*. These snails are amphibious and live on moist soil and vegetation of ditches, slow-moving steams, impoundments, and similar bodies of water. Thus, rice fields are an ideal environment for these hosts. The parasites develop in the snail, and eventually great numbers of infective cercariae are liberated by the snail into the water. Man acquires the infection by contact with the water in which the cercariae are swimming.

The association between schistosomiasis and rice cultivation is seen especially in Leyte and Samar in the Philippines. Rice fields are continuously inundated, and at harvest only the heads are cut from the stems by workers wading barefoot in the fields. The stubble is left behind for fertilizer. Thus, at all times the fields provide ideal conditions for *O. quadrasi*, and workers are continually exposed. There are no sanitary facilities for the disposal of feces; feces from infected persons are deposited in the fields or in connecting ditches.

In mainland China wet rice cultivation also provides one of the chief means of infection. While the snail intermediate host *O. hupensis* does not ordinarily inhabit rice fields, it finds ideal conditions in natural man-made irrigation canals and on low marshy sites that in many parts of the endemic areas are flooded during the wet season of the year. Since this has been accompanied, at least in the past, with the extensive use of night-soil as a fertilizer, the conditions have been highly favorable for transmission. Little information is available on the current status of the problem.

The government of the Philippines, the World Health Organization, and the United States Agency for International Development have conducted a number of control trials, based on elimination of the snail breeding sites, that appear to have had a considerable degree of success (Pesigan et al. 1958). According to McMullen (1973), it is also essential that where schistosomiasis is a problem an effort be made to cultivate rice varieties requiring relatively less water. He also maintains that the practice of planting or sowing rice in rows simplifies weed control, increases yield, and that the attendant cultivation tends to eliminate *Oncomelania* in the fields. Experiments in Japan (McMullen 1973) have shown that under certain conditions water application for rice culture can be reduced as much as 50 percent without affecting yield and that this reduction can prove an effective means of reducing snail breeding in the rice fields and adjacent drains and canals. In the Philippines the number of snails dropped from 200 to less than 1 per square meter, and the rice yield increased by more than 50 percent after the adoption of intermittent application of water.

Whether such yields will be obtained elsewhere is, of course, the crux of the question. As McMullen (1973) notes, the effectiveness of such measures as water management for snail control depends to a great extent on their acceptance and use by the farming community.

SUMMARY

The green revolution has not caused sweeping, all-encompassing changes in the way that rice is grown in Asia. Still, the potential is there for far more extensive changes than have already occurred. In the future, these changes will undoubtedly occur at a slower rate than before because of the factors limiting the spread of green revolution technology.

Where large areas have been newly irrigated for rice production, and where insecticides have been introduced for the control of rice insects, no drastic changes in patterns of disease transmission appear to have occurred. This is undoubtedly in part the result of the relatively small number of disease vectors that are associated with rice fields.

Those vectors that do use rice fields as habitats have been identified. There are only a few areas (Java, South China) in Asia where rice-field *Anopheles* are primary malaria vectors, and a few areas where they play a secondary role. The picture for filariasis is less clear, but rice-field mosquitoes appear to be relatively unimportant except on the west coast of Malaysia--and that focus has contracted sharply as a result of anit-*Anopheles* campaigns. The *Culex* vectors of Japanese encephalitis virus are highly associated with rice culture in many parts of Asia, so much so that there is some evidence, from Korea, that insecticide treatment of rice pests may dramatically reduce the incidence of the disease, at least temporarily. The snail hosts of *Schistosoma japonica* have a wide range of habitats, but in at least some areas (Philippines, China) rice culture may play a significant role in the ecology of the disease.

Where rice culture is a significant feature of the countryside, any change in production technology, or in the extent of fields, may change the habitats of the disease vectors discussed above. In most of Asia, rice production has not been "modernized" by the introduction of agricultural chemicals, large-scale water management technology, and heavy machinery. There is still a potential for the development of sound, integrated pest control programs that would take into account both agricultural pests and vectors of disease.

REFERENCES

Chen, Y. K., H. I. Ree, H. K. Hong, and C. Y. Chow (1967) Bionomics and vector efficiency of *Anopheles sinensis* in Korea. WHO/Mal/67.633.

Chow, C. Y. (1949) The identification and distribution of Chinese anopheline mosquitoes. J. Nat. Malar. Soc. 8: 121-131.

Chow, C. Y. (1950) Mosquito studies in China, past and present. Mosq. News 10:134-137.

Chow, C. Y., R. M. Ibnoe, and S. T. Josopoero (1960) Bionomics of *Anopheline* mosquitoes in inland areas of Java, with special reference to *Anopheles acronitus* Don. Bull. Entomol. Res. 50:647-660.

Covel, G. (1944) Notes on the distribution, breeding places, adult habits, and relation to malaria of the Anopheline mosquitoes of India and the Far East. J. Malar. Inst. India 5:399-434.

Dalrymple, D. G., and W. I. Jones (1973) Evaluating the Green Revolution. Washington, D.C.: Bureau for

Program and Policy Coordination, U.S. Agency for International Development.

Dalrymple, D. G. (1974) Development and Spread of High-Yielding Varieties of Wheat and Rice in the Less Developed Nations. U.S. Department of Agriculture, Foreign Development Division, Economic Research Service. FAER No. 95.

Detels, R., M. D. Cates, J. H. Cross, G. S. Irving, and R. H. Watten (1970) Ecology of Japanese encephalitis virus on Taiwan in 1968. Am. J. Trop. Med. Hyg. 19: 716-723.

Gould, D. J., H. C. Barnett, and W. Suyemoto (1962) Transmission of Japanese encephalitis virus by *Culex gelidus* Theobald. Trans. Roy. Soc. Trop. Med. Hyg. 56:429-435.

Gould, D. J., S. Esah, and V. Pranith (1967) Relation of *Anopheles aconitus* to malaria transmission in the central plain of Thailand. Trans. Roy. Soc. Trop. Med. Hyg. 61:441-442.

Gresser, J., J. L. Hardy, S. M. K. Hu, and W. F. Scherer (1958) The growth curve of Japanese encephalitis virus in the vector mosquito of Japan, *Culex tritaeniorhynchus*. Jap. J. Exp. Med. 28:243-248.

Griffin, K. (1972) The Green Revolution: An Economic Analysis. Geneva: U.N. Research Institute for Social Development.

Hardenburg, W. E. (1922) Mosquito Eradication. New York: McGraw-Hill Book Company.

Harrison, B. A. (1972) A new interpretation of affinities within the *Anopheles hyrcanus* complex of Southeast Asia. Mosq. Syst. 4:73-83.

Hu, S. M. K. (1935) Notes on the relative adult density of *Anopheles hyrcanus* var. *sinensis* Wiedemann during 1933 with reference to malaria incidence in Kaochiao Shanghai area. Chin. Med. J. 49:469-474.

Hu, S. M. K., and H. Yu (1937) Further studies on the blood preferences of *Anopheles hyrcanua* var. *sinensis* in Shanghai region. Chin. Med. J. 51:639-642.

International Rice Research Institute (1971a) Current Insect Problems and Activities in Indonesia. IRRI Conference, April 1971. Mimeo. Manila.

International Rice Research Institute (1971b) Rice Policy Conference: Current Papers from the Department of Agricultural Economics. Los Baños.

International Rice Research Institute (1974) Research Highlights for 1973. Los Baños.

MacDonald, W. W., C. E. G. Smith, P. S. Dawson, A. Ganapathipillai, and S. Mahadevan (1967) Arbovirus infections in Sarawak: Further observations on mosquitoes. J. Med. Entomol. 4:146-157.

McMullen, D. B. (1973) Biological and environmental control of snails, pp. 533-591. *In* Epidemiology and Control of Schistosomiasis, N. Ansari (ed.). Baltimore: University Park Press.

Moorehouse, D. E. (1962) Notes on the bionomics of *Anopheles campestris* Reid, and on its disappearance following house-spraying with residual insecticides. Med. J. Malaya 18(3):184-192.

Muangman, D., R. Edelman, M. J. Sullivan, and D. J. Gould (1973) Experimental transmission of Japanese encephalitis virus by *Culex fuscocephala*. Am. J. Trop. Med. Hyg. 21(4):482-486.

Neogy, B. P., and A. K. Sen (1962) *Anopheles stephensi* as a carrier in rural Bengal. Ind. J. Malar. 16:81-85.

Pesigan, T. P., N. G. Hairston, J. J. Jauregui, E. G. Garcia, A. T. Santos, B. C. Santos, and A. A. Besa (1958) Studies on Schistosoma japonicum infection in the Philippines. Bull. WHO 18:481-578.

Rao, V. V. (1949) Malaria in Orissa. Ind. J. Malar. 3:151-163.

Reid, J. A. (1953) The *Anopheles hyrcanus* group in Southeast Asia (Diptera: Culicidae). Bull. Entomol. Res. 44:5-76.

Reid, J. A. (1962) The *Anopheles barbirostris* group (Diptera: Culicidae). Bull. Entomol. Res. 58:1-57.

Scanlon, J. E., E. L. Peyton, and D. J. Gould (1968) An annotated checklist of the Anopheles of Thailand. Thai National Scientific Papers, Fauna Series, No. 2. Bangkok: Thailand. Applied Scientific Research Corp.

Timmer, C. P., and W. P. Falcon (1975) The political economy of rice production and trade in Asia. *In* Agriculture in Development Theory, Lloyd G. Reynolds (ed.). New Haven: Yale University Press. p. 376.

Viswanathan, D. K., Sribas Das, and A. V. Oommen (1941) Malaria-carrying anophelines in Assam, with special reference to the results of twelve months' dissections. J. Malar. Inst. Ind. 4:297-306.

Wada, Y., S. Kawai, S. Ito, T. Oda, J. Nishigaki, N. Omori, K. Hayashi, K. Mifune, and A. Schichijo (1967) Ecology of vector mosquitoes of Japanese encephalitis, especially *Culex tritaeniorhynchus*. Trop. Med. 9:45-57.

Wharton, C. R., Jr. (1969) The green revolution: Cornucopia or Pandora's box. Foreign Affairs 47:464-476.

Wharton, R. H., A. B. G. Laing, and W. H. Cheong (1964) Studies on the distribution and transmission of malaria and filariasis among aborigines in Malaya. Ann. Trop. Med. Parasitol. 57:235-254.

Wilson, R. (1961) Filariasis in Malaya--A general review. Trans. Roy. Soc. Trop. Med. Hyg. 55:107-129.

APPENDIX B: ACRONYMS USED IN THIS REPORT

AFPCB	Armed Forces Pest Control Board
AID	Agency for International Development
AMCA	American Mosquito Control Association
CDC	Center for Disease Control, Atlanta
CSIRO	Commonwealth Scientific and Industrial Research Organization
DANIDA	Danish International Development Agency, Copenhagen
FAO	Food and Agriculture Organization of the United Nations, Rome
ICAITI	Instituto Centroamericano de Investigacíon y Technologí́a Industrial, Guatemala City
ICIPE	International Centre for Insect Physiology and Ecology, Nairobi
ICMR	India Council for Medical Research, New Delhi
IDRC	International Development Research Center, Ottawa
INCAP	Instituto Centroamericano de Plaguicidas
IRRI	International Rice Research Institute, Los Baños, The Philippines
NPCA	National Pest Control Association
OIRSA	Organismo Internacional Regional de Sanidad Agropecuaria, San Salvador
ORSTOM	Office de la Recherche Scientifique et Technique Outre-Mer, Paris
PAHO	Pan American Health Organization, Washington
PCO	Pest Control Operation
SIDA	Swedish International Development Authority, Stockholm
UNDP	United Nations Development Program

UNICEF	United Nations Children's Fund
UNIDO	United Nations Industrial Development Organization, Vienna
USPHS	United States Public Health Service
WHO	World Health Organization of the United Nations, Geneva

APPENDIX C: CHEMICAL FORMULAE OF PESTICIDES MENTIONED IN THIS REPORT

Abate	*See* temephos
aldrin	Not less than 95 percent of 1,2,3,4,10,10-hexachloro-1,4,4a,5,8,8a-hexahydro-1,4-*endo-exo*-5,8-dimethanonaphthalene
allethrin	2,2-dimethyl-3-(2-methyl propenyl) cyclopropane carboxylic acid ester with 2-allyl-4-hydroxy-3-methyl-2-cyclopenten-1-one
Altosid	*See* methoprene
apholate	2,2,4,4,6,6-hexakis(1-aziridinyl)-2,2,4,4,6-hexahydro-1,3,5,3,4,6-triazatriphosphorine
Baygon	*See* propoxur
Baytex	*See* fenthion
BHC	*See* HCH
bromophos	O-(4-bromo-2,5-dichloro phenyl) O,O-dimethyl phosphorothioate
BUX	*m*-(1-ethylpropyl)phenyl methylcarbamate mixture (1-4) with *m* (1-methylbutyl) phenyl methylcarbamate
carbaryl	1-naphthyl methylcarbamate
carbofuran	2,3-dihydro-2,2-dimethyl-7-benzofuranyl methylcarbamate
chlorphoxim	ortho-chlorophenylglyoxylonitrile oxime with O,O-diethyl phosphorothioate
chlorpyrifos	O,O-diethyl-O-(3,5,6-trichloro-2-pyridyl) phosphorothioate
chlorpyrifos methyl	O,O-dimethyl-O-(3,5,6-trichloro-2-pyridyl) phosphorothioate
DDT	1,1,1-trichloro-2,2-bix (p-chlorophenyl) ethane

deet	N-N-diethyl-*m*-toluamide
Dibrom	*See* naled
dieldrin	Not less than 85 percent of 1,2,3, 4,10-10-hexachloro-6-7-epoxy-1, 4,4a,5,6,7,8,8a-*endo*-*exo*-5-8 dimethanonaphthalene
dimethoate	0,0-dimethyl S-(methyl carbamoyl-methyl) phosphorodithioate
dimethrin	2,4-dimethylbenzyl 2,2-dimethyl-3-(2-methylpropenyl) cyclopropane carboxylate
diphacinone	2-diphenylacetylindan-1,3-dione
Dursban	*See* chlorpyrifos
EPN	0-ethyl 0-*p*-nitrophenylphenylphos-phonothioate
fenitrothion	0,0-dimethyl 0-(4-nitro-*m*-tolyl) phosphorothioate
fensulfothion	0,0,-diethyl 0-*p*[(methyl-sulfinyl) phenyl]phosphorothioate
fenthion	0,0-dimethyl 0-[3-methyl-4-(methyl-thio)phenyl] phosphorothioate
Flit MLO	Petroleum distillate of undisclosed nature
fumarin	3(*a*-acetonylfurfuryl)-4-hydroxy-coumarin
gamma HCH	*See* lindane
HCH	1,2,3,4,5,6-hexachlorocyclohexane, mixed isomers
indalone	Dimethyl phthalate *n*-butyl mesityl oxide oxalate
Iodofenphos	0-(2,5-dichloro-4-iodophenyl) *0,0*-dimethyl phosphorothioate
Dimirin	3,4,5-trimethylphenyl methylcarba-mate, 75 percent; 2,3,5-trimethyl-phenyl methylcarbamate, 18 per-cent
Largon	1-(4-chlorophenyl),3-(2,6-difluoro-benzoyl) urea
lindane	1,2,3,4,5,6-hexachlorocyclohexane, 99 percent or more gamma isomer
London purple	A mixture of calcium arsenite, calcium arsenate, and miscella-neous materials, including a small amount of dye
malathion	diethyl mercaptosuccinate, S-ester with 0,0-dimethyl phosphorodi-thioate

methoprene	isopropyl (2E,4E)-11-methoxy-3,7, 11,-trimethyl-2,4-dodecadienoate
methoxychlor	1,1,1-trichloro-2,2-bix (*p*-methoxy-phenyl) ethane
methyl Dursban	*See* chlorpyrifos methyl
metepa	tris(2-methyl-1-aziridinyl) phos-phine oxide
methyl parathion	0,0-dimethyl 0-*p*-nitrophenyl phos-phorothioate
naled	1,2-dibromo-2,2-dichloroethyl di-methyl phosphate
parathion	0,0-diethyl 0-*p*-nitrophenyl phos-phorothioate
Paris green	copper acetoarsenite $(CH_3COO)_2$ $Cu.3Cu\ (AsO_2)_2$
phenthoate	ethyl mercaptophenylacetate 0,0-dimethyl phosphorodithioate
phoxim	phenylglyoxy[illegible]e oxime 0-ester with 0,0-diethyl phosphorothioate
pindone	2-pivaloylindane-1,3-dione
piperonyl butoxide	α-[2-(butoxyethoxy) ethoxy]-4,5-methylenedioxy-2-propyltolune
propoxur	0-isopropoxyphenyl methylcarbamate
pyrethrins	Mixture of four compounds from the pyrethrum plant, *Chrysanthemum cinariaefolium*
red squill	The powdered bulbs of red squill, *Urginea* (*Scilla*) *maritima*
resmethrin	(5-benzyl-3-furyl) methyl 2,2-di-methyl-3-(2-methyl propenyl) cyclopropanecarboxylate
Rutgers 612	ethyl hexanediol
sodium fluoacetate	$CH_2F\text{-}COONa$
strychnine	Extract of the alkaloids of nux vomica, *Strychnos* spp.
temephos (Abate)	0,0,0',0'-tetramethyl 0,0'-thiodi-p-phenylene phosphorothioate
thiotepa	tris(1-aziridinyl) phosphine sulfide
toxaphene	chlorinated campene containing 67-69 percent chlorine
trichlorfon	dimethyl (2,2,2-trichloro-1-hydroxy ethyl) phosphonate
zinc phosphide	Zn_3P_2
warfarin	3(α-acetonylbenzyl)-4-hydroxy-coumarin